Sustainability Assessment at the 21st Century

Edited by María José Bastante-Ceca, Jose Luis Fuentes-Bargues, Levente Hufnagel, Florin-Constantin Mihai and Corneliu Iatu

Published in London, United Kingdom

IntechOpen

Supporting open minds since 2005

Sustainability Assessment at the 21st Century
http://dx.doi.org/10.5772/intechopen.78105
Edited by María José Bastante-Ceca, Jose Luis Fuentes-Bargues, Levente Hufnagel, Florin-Constantin Mihai and Corneliu Iatu

Contributors
Alexander Tyrsin, Alfia Surina, Alexey Antonov, Christoph Stadel, Florin-Constantin Mihai, Corneliu Iatu, Manuel Vaquero Pineiro, Paola De Salvo, Francesca Giommi, Mihai Dragomir, Daniel Acs, Sorin Popescu, Diana Dragomir, Moshe Gophen, Mohankumar Kumar Muthu, Banuroopa K Kalyanswamy, Mady Mohamed, Tomas Ekvall, María José Bastante-Ceca, Jose Luis Fuentes-Bargues, Levente Hufnagel

First published in London, United Kingdom, 2020 by IntechOpen
IntechOpen is the global imprint of INTECHOPEN LIMITED, registered in England and Wales, registration number: 11086078, 7th floor, 10 Lower Thames Street, London, EC3R 6AF, United Kingdom
Printed in Croatia

British Library Cataloguing-in-Publication Data
A catalogue record for this book is available from the British Library

Additional hard and PDF copies can be obtained from orders@intechopen.com

Sustainability Assessment at the 21st Century
Edited by María José Bastante-Ceca, Jose Luis Fuentes-Bargues, Levente Hufnagel, Florin-Constantin Mihai and Corneliu Iatu
p. cm.
Print ISBN 978-1-78984-976-9
Online ISBN 978-1-78984-977-6
eBook (PDF) ISBN 978-1-83880-079-6

For EU product safety concerns:
IN TECH d.o.o., Prolaz Marije Krucifikse Kozulić 3, 51000 Rijeka, Croatia, info@intechopen.com or visit our website at intechopen.com.

Meet the editors

María José Bastante-Ceca obtained her PhD in Innovation and Engineering Projects from Universitat Politècnica de València (UPV), Spain in July 2006. Previously, she obtained her degree in Agronomical Engineering in July 1997, and her degree in Environmental Sciences in October 2000.

In 2002, she began her career as a researcher in the Department of Engineering Projects, UPV, where she became a teacher in October 2003, teaching subjects related to environmental impact assessment, eco-design, carbon footprint, and life cycle assessment, among others. In 2012, she was appointed as Assistant Director for the Research Department. She is certified as a Project Manager Associate, IPMA Level D, from 2012, and she has been on the Board of Spanish Association of Project Management and Engineering since July 2012.

Dr Jose Luis Fuentes-Bargues obtained his doctorate in Construction and Production Engineering Program from the National University of Distance Education (UNED), in April 2013. Previously he obtained his degree in Industrial Engineering in September 1999, and his second degree in Technical Architecture in July 2012 from the Universitat Politècnica de València (UPV). He began his career in 2006 as a professor in the Department of Engineering Projects, UPV. Here he taught subjects related to projects, environmental impact assessment, and project management. He taught classes as a professor in the Chemical Engineering Department of the Universitat de València between 2007 and 2016, in the degrees of Chemical Engineering and Environmental Sciences and in the program Quality, Energy, Environmental and Risks Management of the UNED since 2016.

Dr Levente Hufnagel is an associate professor at Szent István University, Gödöllő, Hungary, in climate change and ecosystem ecology, biogeography, biometrics, and ecological research methodology. He has over 20 years of experience at leading Hungarian academic institutions teaching PhD, MSc, and BSc students from various social and cultural backgrounds. He has more than 160 scientific publications (in both aquatic and terrestrial ecological aspects of plants, animals and microbes, at community as well as population level) and more than 700 independent citations. He has contributed as a participant of several big ecological research and development projects, having significant experience in multidisciplinary co-operations (with more than 180 co-authors in different publications). He has been the supervisor of several PhD, BSc, and MSc theses, and an editor-in-chief of an international scientific journal indexed by Web of Science and Scopus. He has wide supervising and editing experience. Dr Hufnagel graduated from Eötvös Lorand University with his Master's degree in Ecology and Evolutionary Biology and his PhD degree in Hydrobiology. He also graduated from Szent István University with a PhD degree in Agricultural Science and with other degrees from Corvinus University of Budapest and Adventist Theological College.

Florin-Constantin Mihai holds a PhD degree in Geography (2013) with maximum distinction from the Department of Geography, "Alexandru Ioan Cuza" University of Iasi (Romania), and BSc and MSc degrees in Environmental Science. He has published papers in international peer-reviewed journals, conference proceedings, and book chapters on various topics regarding environmental and waste management issues. His research aims to develop new methods and indicators to assess key environmental issues across various geographical scales. His research interests include environmental geography, waste management, pollution, sanitation, sustainability, circular economy, public and environmental policies, spatial analysis, and rural and regional development. He has previously edited the following books: "E-waste in Transition: from Pollution to Resource", "Solid Waste Management in Rural Areas", and "Environmental Risks". These are available at https://www.intechopen.com/.

Corneliu IATU is a Full Professor at Alexandru Ioan Cuza University of Iasi (the oldest Romanian university) with specializations in human geography and territorial planning. He is the President of the Romanian Geographical Society, a member of the Europea Academy and Rector Deputy of Strategy, Institutional Development and Quality Management of Alexandru Ioan Cuza University. He has published more than 20 books including under foreign publishing houses (France), and also translations of books such as Introducing à la géographie humaine by. A. Bailly, H. Beguin, and R. Scariati. His research interests are economic geography, rural geography, political geography, and sustainable development. He has supervised several Ph.D theses on such topics, some of them in partnership with French universities (Tours, Angers, Bordeaux, Paris IV Sorbonne etc.).

Contents

Preface

Since the second half of the 20th century, the big processes of globalization of the economy, coupled with the development of new technologies and the increase of the population, have led to the emergence of major environmental problems whose importance transcends the limits of the countries, becoming global environmental impacts.

Sustainability is one of the most important challenges for the societies of the 21st century. The conservation of the planet, as well as the fight against climate change, are crucial for the next generations to maintain a liveable world.

This book presents the current state of sustainability and intends to provide the reader with a critical perspective of how the 21st century societies must change their development model facing the new challenges (globalization, internet of things, industry 4.0, smart cities, circular economy, sustainable agriculture and so on), in order to achieve the Sustainable Development Goals of Agenda 2030.

The first section presents a short introduction to topics like sustainability, circular economy and sustainable development goals.

The second section covers theories about sustainable development presenting a new approach for this concept based on probabilistic and entropy theories.

The third section presents some different methodologies about sustainability assessment applied to different fields of research such as water management, software development, life cycle assessment, or smart and sustainable agriculture.

Finally, the fourth section covers different case studies varying from the presentation of a sustainability assessment tool to green building, the presentation of a case study of rural tourism and territorial development in Italy and the management of vertical and horizontal archipelagos of agriculture and rural development in the Andean realm, among others.

To end, we want to thank all the authors who contributed to this book, since they provide different approaches and very valuable contributions to face this topic from different points of view with the common objective of making the world a better place to live.

María José Bastante-Ceca and José Luis Fuentes-Bargues
Universitat Politècnica de València,
GIDDP, Departamento de Proyectos de Ingeniería,
Valencia, Spain

Florin-Constantin Mihai
Alexandru Ioan Cuza University,
Department of Research, Faculty of Geography and Geology,
Iasi, Romania

Corneliu Iatu
Alexandru Ioan Cuza University,
Department of Geography, Faculty of Geography and Geology,
Iasi, Romania

Levente Hufnagel
Szent István University,
Laboratory of Biometrics and Quantitative Ecology, Institute of Crop Production,
Faculty of Agricultural and Environmental Sciences,
Gödöllő, Hungary

Section 1

Introduction

Chapter 1

Introductory Chapter: The Need to Change the Paradigm - Sustainability and Development at the 21st Century

María José Bastante-Ceca, José Luis Fuentes-Bargues, Mihai Florin-Constantin, Corneliu Iatu and Levente Hufnagel

1. Introduction

Since the second half of the twentieth century, the big processes of globalization of the economy, coupled with the development of new technologies and the increase of the population, have led to the emergence of major environmental problems whose importance transcends beyond the limits of the countries, in a manner that we could say that they are global impacts. These problems include, among others, the ozone layer depletion, the climate change due to the greenhouse gas emission, or the depletion of natural resources.

Industry, as well as modern societies, must face this challenge, changing their consumption patterns, increasing product life, banishing the concept of "use and throw away," and changing from the traditional productive systems to a more sustainable ones.

Sustainable Development's most recognized definition was stablished at the publication *Our Common Future*, known as *Brundtland Report* [1], as:

"development that meets the needs of the present without compromising the ability of future generations to meet their own needs. It contains within it two key concepts:

- The concept of 'needs', in particular, the essential needs of the world's poor, to which overriding priority should be given; and
- The idea of limitations imposed by the state of technology and social organization on the environment's ability to meet present and future needs."

Traditionally, Sustainable Development concept has been symbolized as three circles representing the triple bottom line of sustainability: society, economy, and environment. Nevertheless, different authors have proposed alternative representations, to consolidate the concept of society, environment, and economy as pillars of the sustainability, as can be seen in **Figure 1**.

In 2015, the United Nations adopted the 2030 Agenda for Sustainable Development, which went into force on 1 January 2016, after the Paris Agreement was adopted at UN Conference of Climate Change held in Paris in November 2015. This Agenda includes the 17 Sustainable Development Goals (SDGs), which have the aim to end poverty, to fight inequality, and to help countries to promote prosperity while preserving the environment at the same time (**Figure 2**).

IntechOpen

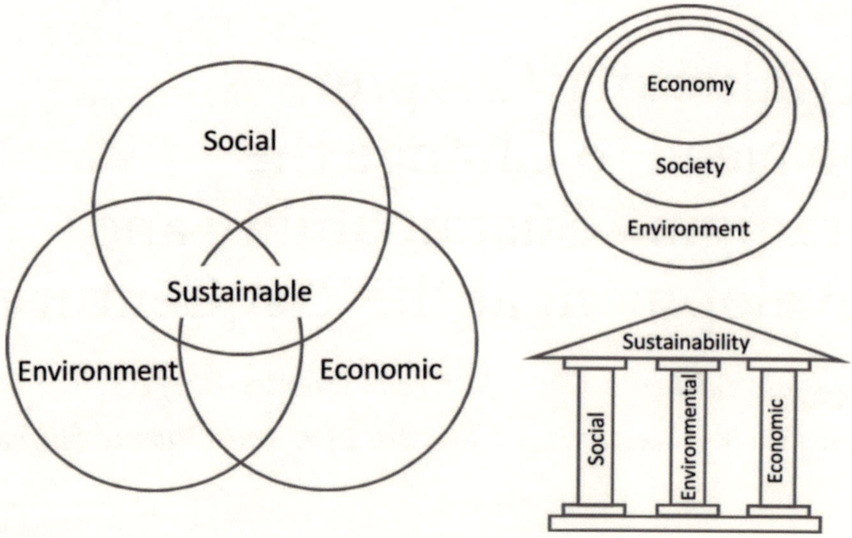

Figure 1.
Different representations of sustainability concept: as three intersecting circle, literal "pillars" and a concentric circles approach [2].

Figure 2.
The 17 sustainable development goals [3].

The UN SDGs aim ambitions and necessary targets across a wide range of socio-economic, environmental, and governance issues in order to reduce the significant gaps between high-income countries and emerging economies in terms of population access to critical services (health, education, public utilities, infrastructure) and to limit the extreme poverty among the most vulnerable populations from Asia, Africa, Latin America, or Eastern Europe.

By the twenty-first century, humankind has fallen into a very complex global human-ecological crisis which endangers not only its economic system, general welfare, peace, and development but its long-term survival and mere existence as well. This crisis requires effective international action and coordinated joint work, but the humankind has reached this time torn to 195 independent national states without having an authorized global organization or effective cooperation system which would represent common interests of humankind efficiently. Natural processes are basically global, since climate change, overpopulation, and contamination of oceans, rivers, and atmosphere do not know state borders.

Sustainable society is an aimed global way of cooperation which ensures the survival of humanity, the constant preservation of our living conditions, the protection of the regulation capacity of the biosphere and its high biodiversity (as the guarantee of reliable natural operation), the good operation of the global economic system, the reduction of social tensions (e.g., inequality, famine, extreme poverty, crime, riots, terrorism, aggression, wars), the scientific and technological development, as well as the preservation and development of our natural and cultural heritage in the long term. The establishment of a sustainable society depends on macro-level (law, political will, consensus, public support) and micro-level conditions (affecting the everyday operation of individuals, families, companies, and small communities). When scientists make an effort in order to save, for example, an endangered species [4, 5], they might not consider this abovementioned complexity of the whole problem, which would make the work necessary.

The sustainability of the human society is endangered by the global human-ecological crisis and a lot of global problems, which are in close relationship with each other. In this phenomenon, the global population explosion (overpopulation of our planet) has a central role, because more people have a larger ecological footprint, a larger consumption, and more intensive pollution, occupy more space from natural ecosystems, and emit more carbon dioxide through their activities of course.

At the same time, higher population density directly enhances aggression (crime, riots, revolutions, demonstrations, wars, and terrorism) and the risk of public health problems, epidemics, pandemics, and the change of land use [6]. Climate change results in significant transformation of the biosphere and biological diversity pattern of the Earth [7, 8]. Biodiversity crisis (extinction of key species and the reduction of habitats) and climate change induce each other in a positive feedback loop, since through the biosphere, climate-regulating ecosystem services are weakened. Overpopulation and social crisis are in a similar positive feedback loop, since it is proven that poverty and hopelessness increase the number of offspring. People living in extreme poverty have nothing to distribute and nothing to base the future on; that is why many of them change from “K” to “r” reproduction strategy, trusting that some of their offspring will survive. Social crisis and public health crisis as well as social crisis and aggression (violence, crime, terrorism, riots, and civil war) are in a similar feedback loop.

2. The role of circular economy in achieving SDGs

Linear economy (“take-make-consume-dispose“) has significant limitations in terms of sustainability through exploitation of natural resources, destroying natural ecosystems and promoting excessive consumption patterns while generating huge amounts of solid waste and wastewater (municipal, industrial, and agricultural sources) which pollute environment through illegal waste dumping sites, landfills, incinerators, and lack or poor wastewater treatment plants.

Economic growth must be provided based on sustainability main pillars such as economic, social, and environmental nexus in a multi-scale context from global standards toward regional and local levels. Future predictions show that population growth and rural–urban migration will emerge in these regions and human pressures on environment in terms of energy and water supply demands, agricultural land reclamation, urbanization process, biodiversity loss, waste production, and plastic pollution are likely to increase to alarming levels. Both developed and emerging economies must cope with effects of climate change, water shortage risks, industrial pollution, food security, demographic challenges, and socioeconomic inequalities. In this context, linear economy is clearly unsustainable in the medium and long terms, and shifting transition toward circular economy is quite necessary. This type of economy aims to cut as much as possible the natural resource depletion through reusing of secondary materials and closing the production and consumption loops by avoiding further waste generation and their disposal in landfills. In fact, waste management sector must be replaced by resource management (e.g., "end of waste"), and in this regard, a new paradigm is born such as "zero waste cities" as ultimate sustainability goals.

The 3R policy (reduce-reuse-recycle) based on waste hierarchy concept (where landfill and waste incineration are regarded as the least favorable options) is supplemented by product life expansion alternatives (repair-recovery-refurbish-repurpose-remanufacture) and to rethink our consumption patterns and to refuse to buy nonrecyclable items. These actions are more suitable than material recycling where additional raw materials and energy are needed for making new products. However, secondary materials and renewable energy sources should feed the new circular economy system instead of raw materials and fossil fuels. Composting of biowaste fraction or anaerobic digestion must be used to produce organic fertilizers and biofuels instead to be landfilled. Upcycling or creative reuse and sharing economy are other mechanisms that lead to responsible consumption patterns. The circular economy is strongly related to SDG9, industry, innovation, and infrastructure; SDG11, sustainable cities and communities; and SDG12—responsible production and consumption. Also, circular economy is interconnected to green economy (promoting clean energy sources, sustainable waste management practices, organic agriculture, etc.) or blue economy (sustainable management of marine resources and conservation). Full access of the population (urban and rural) to improved sanitation, solid waste, and wastewater management practices is critical to fulfilling SDG3 (good health and well-being), SDG6 (clean water and sanitation), and SDG10 (reducing inequalities) and to be able to make transition from linear to circular economy.

New digital technologies and Internet networks provide new tools for urban areas to increase their resource efficiency and reduce their ecological footprint that is becoming the so-called smart cities. A key aspect is to use the "Internet of things" (IoT) and big data to manage future megacities in a sustainable manner. According to Ellen McArthur foundation, "A circular economy is based on the principles of designing out waste and pollution, keeping products and materials in use, and regenerating natural systems" [9]. In other words, circular economy plays a key role in sustainability of cities and rural communities. EU policies recognize the key role of future circular economy in Europe by dedicating a special package document [10] . This will enforce waste reduction targets and high rates for packaging materials such as plastic, glass, paper and cardboard, wood, aluminum, and ferrous materials with specific deadlines for 2025 and 2030. The ultimate goal is to reach 70% of packaging materials by 2030 and recycling 65% of municipal waste stream by 2035 [11]. On the other side, many developing countries must upgrade their waste management infrastructure and increase the collection efficiency in the context of rapid urbanization and demographic explosion expected to happen in Africa and Asia. Developed countries must invest and expand their waste recovery and recycling facilities and stop the export of packaging waste or e-waste items into developing countries (e.g., Malaysia, Indonesia, Ghana, and so on). Urban mining, which focuses on recovery process of valuable materials from used items (e.g.,

DOI: http://dx.doi.org/10.5772/intechopen.90655

e-waste flows), and construction and demolition waste stream have a great potential to feed industry with reliable secondary materials. New product design and production systems and less reliance on packaging materials are required to be enabled across the manufacturing industry. Therefore, the transition of current societies to a circular economy model constitutes a critical pathway toward sustainability.

This book presents a vision of the current state of sustainability and intends to provide the reader with and make a critical perspective of how the twenty-first century societies must change their development model facing the new challenges (globalization, Internet of things, industry 4.0, smart cities, and so on), in order to achieve the SDGs of Agenda 2030.

Author details

María José Bastante-Ceca[1*], José Luis Fuentes-Bargues[1], Mihai Florin-Constantin[2], Corneliu Iatu[3] and Levente Hufnagel[4]

1 GIDDP, Departamento de Proyectos de Ingeniería, Universitat Politècnica de València, Valencia, Spain

2 Department of Research, Faculty of Geography and Geology, "Alexandru Ioan Cuza" University, Iasi, Romania

3 Department of Geography, Faculty of Geography and Geology, "Alexandru Ioan Cuza" University, Iasi, Romania

4 Faculty of Agricultural and Environmental Sciences, Laboratory of Biometrics and Quantitative Ecology, Institute of Crop Production, Szent István University, Gödöllő, Hungary

*Address all correspondence to: mabasce1@dpi.upv.es

References

[1] Brundtland GH. Our Common Future. USA: Oxford University Press; 1990. 420p

[2] Purvis B, Mao Y, Robinson D. Three pillars of sustainability: in search of conceptual origins. Sustainability Science. 2019;**14**:681-695. DOI: 10.1007/s11625-018-0627-5

[3] The 17 Sustainable Development Goals [Internet]. 2019. Available from: http://sustainabledevelopment.un.org/sdgs [Accessed: November 04, 2019]

[4] Gilián LD, Bódis J, Eszéki E, Illyés Z, Biró É, Nagy JG. Germination traits of adriatic lizard orchid (*Himantoglossum adriaticum*) in Hungary. Applied Ecology and Environmental Research. 2018;**16**(2):1155-1171

[5] Suel H, Mert A, Yalcinkaya B. Changing potential distribution of gray wolf under climate change in lake district, Turkey. Applied Ecology and Environmental Research. 2018;**16**(5):7129-7137

[6] Rüstemoğlu H, Uğural S. CO_2 emissions in Iran for 1990-2010: A decomposition analysis. Applied Ecology and Environmental Research. 2017;**15**(4):1833-1846

[7] Garamvölgyi Á, Hufnagel L. Impacts of climate change on vegetation distribution No. 1. climate change induced vegetation shifts in the Palearctic region. Applied Ecology and Environmental Research. 2013;**11**(1):79-122

[8] Hufnagel L, Garamvölgyi Á. Impacts of climate change on vegetation distribution No. 2, climate change induced vegetation shifts in the New World. Applied Ecology and Environmental Research. 2014;**12**(2):255-422

[9] Ellen Macarthur Foundation. 2019. What is the circular economy ? Available from: https://www.ellenmacarthurfoundation.org/circular-economy/what-is-the-circular-economy [Accessed: November 02, 2019]

[10] European Commission. 2019. Final Circular Economy Package. Available from: https://ec.europa.eu/environment/circular-economy/index_en.htm. [Accessed: November 02, 2019]

[11] Directive (EU) 2018/851 of the European Parliament and of the Council of 30 May 2018 amending Directive 2008/98/EC on waste. Official Journal of the European Union. L150, 14.6.2019. pp. 109-140

Chapter 2

Sustainable Rural Development under Agenda 2030

Florin-Constantin Mihai and Corneliu Iatu

1. Introduction

The rural environment is a complex system in which the differences in development are evident both at the subnational and international level. The difficulties related to methodological analysis are due to such rural diversity and the partial lack of comparable indicators which lead to the development of objectives and indicators that respond to both national and international needs. Harmonization should be easier in view of the common goal, but policies and strategies do not always provide the required coherence.

The presence of programmatic documents such as the Agenda 2030 reveals a path that can lead to good practices and reliable results even if they do not offer universal or global certainties. Politics at various levels play a decisive role and not always these take the best decisions regarding the rural environment. Thus, there is a diversity of situations, and the application of models is not necessarily a solution because of a wide spectrum of particular conditions at regional and local levels that must be taken into account. However, some mechanisms must be further developed to comply the international sustainable development perspectives to regional and local scales including rural areas.

Agenda 2030 relies on 17 sustainable development goals and 169 targets supported by the United Nations as a global effort to manage current challenges related to poverty, climate, environmental pollution, geographical inequalities, prosperity, peace, and justice [1]. This key strategic document continues the previous Eight Millennium Development Goals (started in 2000) committed to combat poverty, hunger, disease, and illiteracy, to promote gender equality and to ensure environmental sustainability until 2015 [2].

The Paris Agreement aims to undertake ambitious efforts to combat climate change and adapt to its effects among developed and developing countries and to build future clean and climate-resilient communities [3]. The Paris Agreement and the Agenda 2030 are the most ambitious international initiatives so far which address major concerns related to future economic development perspectives combined with societal and environmental sustainability issues. Such actions must take into consideration the huge rural-urban gaps in terms of socioeconomic conditions and reveal the exposure of rural areas to current societal and environmental threats. Despite the rural-urban migration process, rural areas comprise vast geographical regions where a significant population still lives and faces emerging threats associated with climate change, poverty, and lack of critical infrastructure, particularly across developing and transition countries. Reducing geographical and socioeconomic inequalities in terms of basic needs must be a priority at international level. On the other hand, rural lands feed all basic needs of urban areas' (raw materials, energy sources, food supply, water, etc.) additional labor force while preserving the natural habitats of endemic species (flora and fauna) and landscapes (e.g., protected areas). Rural settlements also contribute to the cultural and patrimonial heritage of each region and country. Therefore, sustainable rural development is a complex issue (environment-economic-social nexus) which must be further addressed with the same attention by academics, international bodies, national and local authorities, professionals, and members of civil society as for urban areas.

2. Societal and environmental threats in rural areas

Rural communities are facing several challenges in the context of climate change, land degradation, deforestation, biodiversity loss, and fragmentation of natural habitats, poverty, and geographical isolation. The rural population is more prone to extreme poverty, famine, social exclusion, and environmental injustice, particularly in developing countries from Africa, Asia, and Latin America. Rural communities depend on local geographical conditions (climate, natural resources, landscape, and geographical barriers, socioeconomic conditions, demographic features) to develop agricultural, industrial, or tourism activities as economic development pathways. A traditional economy based on subsistence agriculture is still widespread across rural regions of the globe. This type of economy is volatile to natural hazards (extreme weather, flash floods, landslides, erosion, drought) and poor agricultural productivity which translates into famine, extreme poverty, land abandonment, and massive migration. Land use management is a key factor for future rural development perspectives and to find the optimal equilibrium between natural habitats, agricultural lands, and built-up areas. **Figure 1** reveals the emerging societal and environmental threats, sectoral approaches, and synergic effects that must be addressed at subnational levels by each country via regional and local authorities towards rural areas.

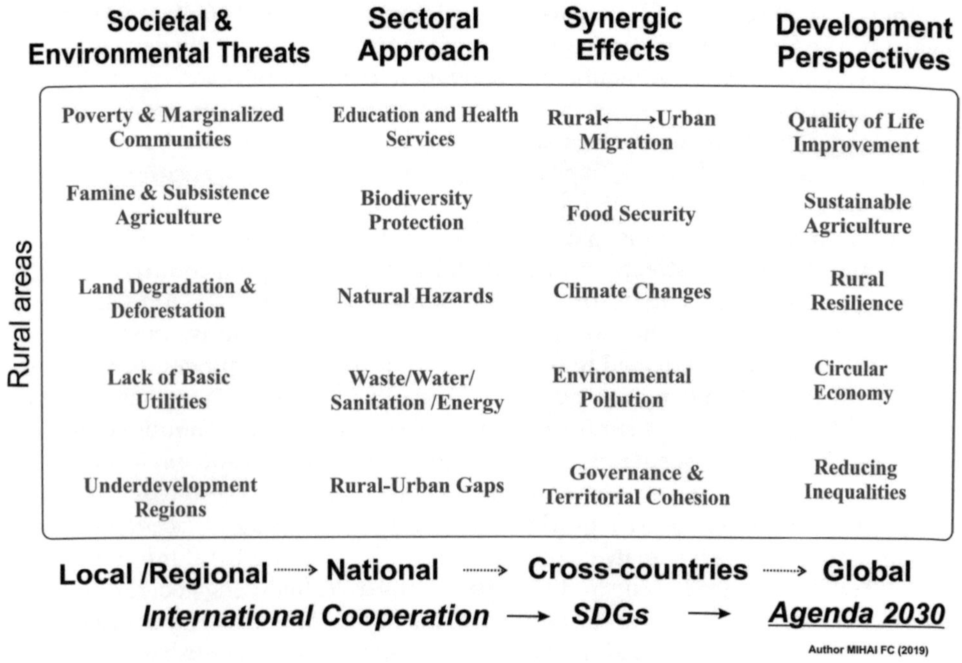

Figure 1.
Challenges of rural communities under agenda 2030 framework.

Rural areas must cope with social, demographic, economic, governance, and environmental challenges. As an example, extensive cattle ranches and emerging oil palm cultivation threaten biodiversity conservation and food security across tropical rural regions while increasing social inequalities and conflicts [4]. On the other hand, agricultural land abandonment (associated with traditional farming, low productivity, poor infrastructure, aging population, massive migration, land ownership change, political instability) has created several socioeconomic and ecological dysfunctionalities in southeastern Europe [5].

Poor agricultural productivity in the Global South is related to the low use of improved seed, use of inappropriate fertilizer, inadequate irrigation, and lack of incentives for farmers in the absence of remunerative markets [6]. Extreme poverty, hunger, and undernourishment and rural depopulation are critical issues to be solved across rural Africa besides the poor access to critical amenities (health/education services, sanitation, and water facilities, energy) [7]. Climate changes, land fragmentation, natural resource depletion, political instability, corruption, and conflict areas will further threaten rural areas of developing countries.

In this context, rural resilience and circular economy are key strategic directions to further develop rural economies and reduce socioeconomic inequalities and environmental injustice coupled with access to proper education. A linear economy based on "take-make-dispose" model feed by consumerism society is harmful for the environment and long-term sustainability of urban and rural areas. The EU is aware of the importance of shifting economic model from linear to an ambitious circular economy framework (https://ec.europa.eu/environment/circular-economy/index_en.htm) based on the 3Rs principles such as "reduce-reuse-recycle."

There are other activities which could be integrated resulting 6Rs policy such as the revaluation (of resources), redistribution (of income) and (improve) relations or 9Rs with another three Rs added such as resilience (adaptability), reassessment (scale value) and restructuring (of the economy) (source: https://www.activesustainability.com/sustainable-life/learnsustainability-the-3rs-6rs-and-9rs/).

This new policy needs to be adopted by each EU country including rural areas of Eastern Europe. In countries like Romania, with over 2800 rural municipalities (communes) and other villages included in urban administrative areas, this transition from linear to circular economy could lead to new rural business opportunities based on responsible production and consumption of natural resources (organic farming, agritourism, local niche products, upcycling or creative reuse, etc.) while promoting local traditions and preserving the rural and natural landscapes. International cooperation is needed to successfully achieve the ambitious SDGs until 2030 at the global level. The development perspectives show some critical objectives which cannot be achieved without improvement of rural conditions across each continent.

3. Rural population access to basic public utilities

Poverty and poor infrastructure are the main drivers for underdevelopment and environmental degradation. Rural settlements must have access to basic public utilities to ensure a decent quality of life in areas without significant geographical restrictions.

At the global level, there are huge rural-urban gaps regarding population access to critical amenities such as drinking water, sanitation, electricity, and waste management services, particularly in low- and middle-income countries. Rapid urbanization in developing countries feeds rural-urban migration where poor people have crowed in slum areas without access to urban main public services threatening public health and local environment. Uncontrolled urban expansion towards surrounding rural lands leads to such informal settlements. There are 1.9 billion rural people without access to formal waste management services, and the coverage rate is under 50 among 105 countries [8]. This critical situation translates into million tons of household waste generated and uncollected each year, which leaks into the natural environment via wild dumps, waste dumping in water bodies, or open burning practices. Freshwater ecosystems are often affected by the uncontrolled disposal of waste which further contaminates the downstream water bodies through heavy rains and floods and finally reaches the marine environment. Wildlife is heavily exposed to plastic pollution where rural communities can contribute directly through fishing activities or indirectly as land-based sources via unsound waste management activities.

A study breaks down the rural infrastructure in China in three major categories: facilities for living and production (e.g., drinking water, irrigation, electricity, roads, wastewater treatment, and waste management), development infrastructure (education and healthcare), and environmental infrastructure (clean energy, green housing, and the environmental protection system) [9].

However, **Figure 2** reveals a complex structure of public utilities as essential services for each rural municipality in the world. The lack or poor coverage of public utilities across rural regions leads to environmental degradation via air-water-soil nexus. These are essential services to maintain a decent standard of public health and to protect the natural environment against daily disturbances induced by human needs and economic activities.

The public utilities involve a multi-sectoral approach and a systemic perspective in adopting best current practices which will provide an easier transition to a diverse range of SDGs as shown in **Figure 2**. Thus, the improvement of such public services via sanitation facilities, water, and waste management, clean energy, road networks, and digital technologies will provide the base to build a sustainable development community.

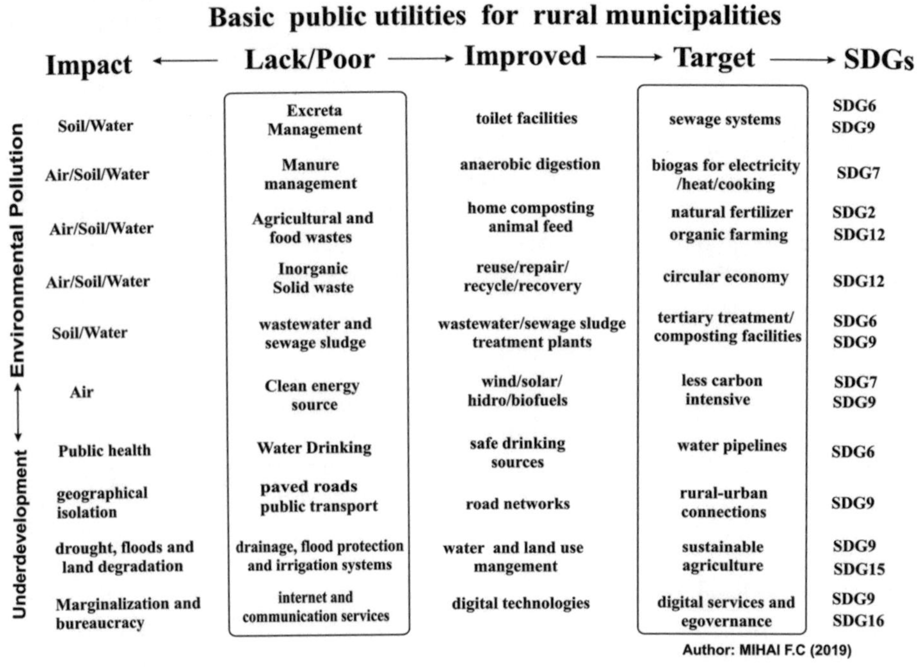

Figure 2.
Public utilities as critical infrastructure for sustainable rural development.

In nowadays, rural areas are still neglected by public services as in the case of former Soviet countries [10]. Furthermore, Central Asia is facing a growing urban-rural divide as a result of a capital city-centric growth model, economic nationalism, and water resource conflicts [11]. Additionally, rural depopulation of Russia makes more difficult to revitalize such regions, despite some recent efforts to improve population access to basic utilities [12]. Poor socioeconomic conditions in rural areas of new EU members stimulate the external migration (abroad) towards older EU countries, as in the case of Romania which facing is labor shortage. Distant rural settlements from urban areas are now facing such labor shortages combined with the aging population process. These peripheral regions lack critical amenities and have poor opportunities for economic development which still depend on traditional farming. Public investments in infrastructure and public utilities are crucial to reconnect such pockets of poverty areas to cities and emerging markets and, on the other hand, to mitigate the rural migration phenomenon.

Improvement of rural mobility by public transport network is essential for access to education (high schools, universities), healthcare services, justice, and social programs.

In developed countries, there is a counter-urbanization process in the proximity of larger cities by movement of people to surrounding rural areas, avoiding noise and air pollution of core cities. There is an intense rural land reclamation for housing, transport, commercial, or industrial infrastructures around "peri-urban" areas including tourism and recreational activities. Therefore, the monopoly of farming activities is replaced by manufacturing, industrial, and various services close to larger urban areas where metropolitan regions emerge. However, this situation is in contrast with distant rural communities with primary focus in exploitation of natural resources and farming activities.

Despite the economic development in the latter years, China must manage several environmental challenges in rural areas such as [9] increasing waste generation rate with insufficient treatment capacity, crude and backward wastewater facilities, regional disparities in terms of sanitation services, and safe drinking-water sources' issues. Similar challenges are facing Indonesia, where rural areas are poorly endowed with infrastructures like roads, sanitation, clean water, and energy [13]:

Biogas is a perfect solution for decentralized off-grid electricity situations in rural areas where an abundance of biowaste is available as feedstock; therefore, biogas could be used for cooking, heating, or gas lighting [14]. Anaerobic digestion is a proper solution to divert organic waste leaking into the natural environment towards an energy source (biogas) of fertilizer (digestate) for agricultural land. Animal feed and home composting practices are additional options to handle the organic waste produced at household level supporting organic farming and animal husbandry. Source separated of dry recyclables (metals, plastics, paper/cardboard, wood) would stimulate recycling and recovery practices via local small- and medium-sized enterprises (SMEs) enacting the first steps towards a rural recycling society.

Rural communities must evaluate the exposure of its territory to natural hazards (floods, heavy rains, heatwaves, hail occurrence, drought, desertification, wildfire) and to take necessary measures to combat such threats. Poor population and peripheral rural areas are most vulnerable to climate change effects due to their reliance on subsistence agriculture.

Rural population access to mobile phones and the Internet could improve agricultural productivity and better land use management practices based on updated knowledge. Also, digital technologies will help rural councils to reduce bureaucracy and increase transparency in community decisions. Better virtual connectivity to high-speed Internet services will provide new collaborative opportunities for rural entrepreneurs including women empowerment.

4. Pathways towards sustainable development goals (SDGs)

The EU Cohesion Policy and Common Agricultural Policy (CAP) are two strategic initiatives which can help to reduce the geographical inequalities in Central and Eastern Europe in terms of basic infrastructure, promoting economic activity and agricultural development and improving the qualifications and skills of the inhabitants, particularly in rural areas [15]. However, the impact of such policies in case of new EU member states needs to be further adjusted with proper funding to boost local and regional economies. The gradual decline of fishery activities across EU rural coastal areas makes it difficult to revitalize these regions, despite new policy incentives such as the Common Fisheries Policy (CFP) as shown in case of Greece [16].

The EU policies and financial instruments must accelerate the mitigation gap between western high-income countries and former Soviet countries of the Eastern Bloc where rural regions are regularly left behind. New projects like LiveRur (https://liverur.eu/) identify the innovative business models that are currently being developed in rural areas based on the sustainable mobilization of resources and better cooperation between operators along the value chain and lead to new services. At regional level, collective forest management supported by small-scale business projects could maintain the network of local produce markets with attractive esthetic values as well as biodiversity conservation [17].

The role of small- and medium-sized enterprises in rural areas is based on local resource use, contributions to the local public budget, job creation, development of infrastructure, and engagement with community [18]. Furthermore, small-scale farmers using agroecological practices can produce the food necessary for diversified, nutritious, sustainable diets, while protecting environmental resources from further degradation [19]. Long-term growth policies should be reoriented to favor small farmers instead of big agribusiness players to maintain food security and social equity in tropical regions [4]. New urban-rural relations, in terms of organic food production, stimulate nearby farmers to adopt the best management practices and to develop nonfarming activities (e.g., tourism and recreational activities, environmental conservation, forest restoration) or urban-rural migration [20].

Rural households that wish to market their products are restricted to local markets, or their production is sold at low prices to intermediaries [12]. This situation is specific to other Eastern European countries where the dispersion of villages, poor road networks, and the urban concentration of services are impediments in the development of direct linkages between local rural producers and urban customers. The development of farmer associations could be a solution in increasing access to regional or even national markets, to provide short supply chains and to reduce reliance on food product imports from abroad, particularly in countries with high potential in agricultural productivity like Romania. Such countries need to raise their rural economies from cheap raw material providers dedicated to exports towards manufactured products and services (e.g., furniture industry, food industry, organic farming, renewable energy, agritourism).

Digital technologies provide new ways to access price and market information, to coordinate input/output resources (including transport and logistics, finance, and production techniques) which could help the agriculture sector in the Global South as shown in several case studies [6].

Improvement of water harvesting, cultivating drought-resistant crops, ecological restoration, combined with better local governance, financial instruments, integrated resource management, sound public services, and better urban-rural linkages could help rural communities around the world to become more sustainable.

Remote rural areas of developing countries should rely on renewable energy sources due to poor coverage of electric grids, high costs of fuel transportation, unsuitable roads, and increasing consumption of biomass fuels with related pollution issues [21].

In this context, "Smart Village" is a promising initiative to provide energy access to remote villages as a catalyst development route for other related sectors such as clean water, sanitation, education, healthcare, and gender equity and support the local markets and democratic engagement as stipulated by SDGs [22].

In poor rural areas of developing countries like Bangladesh, where energy source is based on wood or dried cattle dung, the bioenergy systems (e.g., anaerobic digestion of biowaste) at household level could be a solution in achieving several SDGs [3–5, 7] with societal and environmental benefits despite of major challenges in implementing such projects at large scale associated with severe poverty, poor education, lack of awareness, social and cultural barriers, etc. [23]. Training activities and environmental awareness should combat such barriers, and fortunately, domestic biogas activities start to emerge in developing and transition countries across the globe such as Pakistan, India, China, Vietnam, Laos, Cambodia, Vietnam, Indonesia (Asia); Morocco, Algeria, Cote D'Ivoire, Burkina Faso, Eritrea, Ethiopia, Kenya, Tanzania, Burundi, Rwanda, Uganda (Africa), Colombia, Peru, and Bolivia (South America) [14].

Rural tourism, agritourism, religious tourism, and ecotourism are alternatives or complementary economic activities that could further stimulate rural entrepreneurship while decreasing rural community dependency on one main economic sector (agriculture, forestry, energy, mining, or fishing activities).

Rural communities must respond to wide range of shocks (such as natural events, policy changes, economic disturbances, and insecurity), and successfully managing such risks increases the resilience of a rural community [24]. Sustainable development based on three basic pillars (social, economic, and environmental) could not be achieved without the proper education of the rural population.

The literacy rate is directly proportional to development; thus, full access of rural communities to educational services should be regarded as starting point to achieve ambitious SDGs in developing countries. Also, rural-urban linkages must be addressed as a pathway to stimulate rural development perspectives. These rural-urban dependences may be positive, negative, or neutral. The positive ones are visible especially in developed countries, the negative ones especially in the less developed countries, but neutral relations are difficult to manage particularly in the proximity of urban areas. Regional convergence aims to reduce the geographical inequalities in the distribution of wealth between large cities, towns, and rural municipalities which are part of an administrative region or county. Such approach could strength the urban-rural relations in common projects regarding infrastructure, public services, mobility, business opportunities (e.g., start-up firms, employment growth) and tourism activities involving local stakeholders in community decisions.

5. Conclusions

This chapter draws attention to the societal and environmental threats which rural communities around the world are facing. Agenda 2030 and SDGs aim to eradicate extreme poverty, famine, open defecation, and other critical issues in developing countries associated with lack of public utilities, mainly in rural areas, and to reduce the huge gaps between countries and regions. To achieve all range of SGDs across the globe, proper attention must be paid to rural development perspectives such as quality of life improvement, sustainable agriculture, rural resilience, and circular economy and reduced inequalities. Sustainable rural development involves a holistic approach where daily basic needs of rural populations must be covered by reliable public utilities combined with technical, socioeconomic, and environmental conditions to support regional economies and urban-rural linkages. Rural communities must develop several nonfarming activities coupled with agricultural systems (adapted to local geographical conditions) to become more resilient to economic shocks or environmental disturbances in the context of climate change. Rural areas should receive the same attention and opportunities from decision-makers, academics, and professionals regarding sustainable development policies and investments in infrastructure projects. Agenda 2030 could be achieved if sustainable rural development policies will be implemented in each country next to urban areas.

Author details

Florin-Constantin Mihai[1*] and Corneliu Iatu[2]

1 Department of Research, Faculty of Geography and Geology, "Alexandru Ioan Cuza" University of Iasi, Romania

2 Department of Geography, Faculty of Geography and Geology, "Alexandru Ioan Cuza" University of Iasi, Romania

*Address all correspondence to: mihai.florinconstantin@gmail.com

References

[1] UN 2015. Transforming our world: The 2030 Agenda for Sustainable Development. Resolution adopted by the General Assembly. 2015. Available from: https://www.un.org/ga/search/view_doc.asp?symbol=A/RES/70/1&Lang=E [Accessed: 24 July 2019]

[2] WHO. Millennium Development Goals. Available from: https://www.who.int/topics/millennium_development_goals/about/en/ [Accessed: 27 July 2019]

[3] Adoption of the Paris Agreement. Decision 1/CP.21. Available from: https://unfccc.int/resource/docs/2015/cop21/eng/10a01.pdf [Accessed: 30 July 2019]

[4] Boron V, Payán E, MacMillan D, Tzanopoulos J. Achieving sustainable development in rural areas in Colombia: Future scenarios for biodiversity conservation under land-use change. Land Use Policy. 2016;**59**:27-37. DOI: 10.1016/j.landusepol.2016.08.017

[5] Zakkak S, Radovic A, Nikolov SC, Shumka S, Kakalis L, Kati V. Assessing the effect of agricultural land abandonment on bird communities in southern-eastern Europe. Journal of Environmental Management. 2015;**164**:171-179. DOI: 10.1016/j.jenvman.2015.09.005

[6] Duncombe R, editor. Digital Technologies for Agricultural and Rural Development in the Global South. Oxfordshire, UK: CAB International; 2018. DOI: 10.1079/9781786393364.0000

[7] Khayesi M, editor. Rural Development Planning in Africa. New York, USA: Palgrave Macmillan; 2018. DOI: 10.1057/978-1-349-95297-7

[8] Mihai F-C. One global map but different worlds: Worldwide survey of human access to basic utilities. Human Ecology. 2017;**45**(3):425-429. DOI: 10.1007/s10745-017-9904-7

[9] Zhang X, Li Z. China's rural development road. In: Research Series on the Chinese Dream and China's Development Path. Springer Nature Singapore Pte Ltd. and Social Sciences Academic Press; 2018. DOI: 10.1007/978-981-10-5646-8

[10] Kvasha S, Lesia S, Zhemoyda O. Problems of rural sustainable development in Ukraine. Problems of Agricultural Economics. 2017;**353**(4):125-137 https://doi.org/10.30858/zer/84959

[11] Spoor M. 25 years of rural development in post-soviet Central Asia: Sustaining inequalities. Eastern European Countryside. 2018;**24**(1):63-79. DOI: 10.2478/eec-2018-0004

[12] Wegren S. The quest for rural sustainability in Russia. Sustainability. 2016;**8**(7):602. DOI: 10.3390/su8070602

[13] Sutiyo, Maharjan KL. Decentralization and Rural Development in Indonesia. Springer Nature Singapore Pte Ltd.; 2017. DOI: 10.1007/978-981-10-3208-0

[14] German Biogas Association. Biowaste to Biogas. Production of Energy and Fertilizer from Organic Waste. 2019. Available from: https://www.biowaste-to-biogas.com/ [Accessed: 30 July 2019]

[15] Dudek M, Wrzochalska A. Making development more sustainable? The EU cohesion policy and socio-economic growth of rural regions in Poland. European Journal of Sustainable Development. 2017;**6**(3):189-200. DOI: 10.14207/ejsd.2017.v6n3p189

[16] Loizou E, Chatzitheodoridis F, Polymeros K, Michailidis A, Mattas K. Sustainable development of rural coastal areas: Impacts of a new fisheries policy. Land Use Policy. 2014;**38**:41-47. DOI: 10.1016/j.landusepol.2013.10.017

[17] Lopes A, Diaz-Maroto J. Input of communal forests to the sustainable development of the rural population: Study case of northern Portugal and Galicia. In: Proceedings of International Scientific Conference "Rural Development 2017." Presented at the Rural Development. Lithuania: Aleksandras Stulginskis University; 2018. DOI: 10.15544/RD.2017.227

[18] Kubíčková L, Morávková M, Tuzová M, Nečas I. The role of small and medium-sized enterprises in the development of rural areas. Acta Universitatis Agriculturae et Silviculturae Mendelianae Brunensis. 2017;**65**(6):1987-1996. DOI: 10.11118/actaun201765061987

[19] Anderson MD. Roles of rural areas in sustainable food system transformations. Development. 2015;**58**(2-3):256-262. DOI: 10.1057/s41301-016-0003-7

[20] Silva da RFB, Rodrigues MDA, Vieira SA, Batistella M, Farinaci J. Perspectives for environmental conservation and ecosystem services on coupled rural-urban systems. Perspectives in Ecology and Conservation. 2017;**15**(2):74-81. DOI: 10.1016/j.pecon.2017.05.005

[21] Afsharzade N, Papzan A, Ashjaee M, Delangizan S, Van Passel S, Azadi H. Renewable energy development in rural areas of Iran. Renewable and Sustainable Energy Reviews. 2016;**65**:743-755. DOI: 10.1016/j.rser.2016.07.042

[22] Gevelt T, Canales Holzeis C, Fennell S, Heap B, Holmes J, Hurley Depret M, et al. Achieving universal energy access and rural development through smart villages. Energy for Sustainable Development. 2018;**43**:139-142. DOI: 10.1016/j.esd.2018.01.005

[23] Rahman KM, Edwards DJ, Melville L, El-Gohary H. Implementation of bioenergy systems towards achieving United Nations' sustainable development goals in rural Bangladesh. Sustainability. 2019;**11**(14):3814. DOI: 10.3390/su11143814

[24] Freshwater D. Vulnerability and resilience: Two dimensions of rurality. Sociologia Ruralis. 2015;**55**(4):497-515. DOI: 10.1111/soru.12090

Section 2

Theories about Sustainable Development

Chapter 3

Probabilistic-Entropic Concept of Sustainable Development of the Example of Territories

Alexander Nikolaevich Tyrsin, Al'fiya Adgamovna Surina and Alexey Sergeyevich Antonov

Abstract

Nowadays the problem of sustainable development became one of the topical issues. However, many authors point to conceptual complexity, which is that the concept "sustainable development" includes two terms "development" and "sustainability." At the same time, each of these terms is treated not unambiguously. It leads to the emergence of different interpretations of sustainable development in relation to specific systems. A new concept of sustainable development of systems is proposed. According to this concept, the sustainable development of a complex multidimensional system will be understood as the dynamics consisting in the presence of a trend of balanced change in the entropies of randomness and self-organization while maintaining an acceptable risk level in multidimensional systems. The proposed concept is approved on practical examples. Dynamics of vector entropy and multidimensional risk of Yekaterinburg and Sverdlovsk regions in 1992–2017 is given.

Keywords: sustainable development, differential entropy, risk, multidimensional random variable, monitoring, vector, randomness, self-organization

1. Introduction

The term "sustainable development" originally was used in 1972 at the United Nations Conference on the Human Environment in Stockholm. In 1987 in a report entitled as "Our common future" of the World Commission on Environment and Development (WCED) where Norwegian Prime Minister Gro Harlem Brundtland was a chair, the term definition of "sustainable development" has been formulated: "development that meets the needs of the present without compromising the ability of future generations to meet their own needs" [1]. The "sustainable development" interpretation is very common and does not show a particular way to move into practice [2]. One can point out the main conceptual complexity, which is that the concept of "sustainable development" includes two terms "sustainability" and "development." Moreover, each of these terms is interpreted in different points.

These terms are multidirectional. Really, maximizing efficiency usually increases risks, reducing the stability of the functioning of a system. Rather, excessive stability leads to an increase in the costs of its maintenance, reducing the efficiency of functioning of a system.

Thirdly, formalization of sustainable development is a complicated complexity of the studied systems and the phenomena. Currently, there is no unambiguous, accurate interpretation of a concept of the complex system. However, there are characteristic signs, such as multidimensionality, multiconnectivity, a multiloop, multileveled (hierarchy), the composite and multipurpose nature of construction, and also indeterminacy and stochasticity of behavior. We will give below the definition which, in our opinion, most adequately characterizes the concept of the complex system.

Definition 1. A complex system is called a system in the model of which there is not enough information to effectively manage this system [3].

This fact leads to different understandings of "sustainable development" in relation to particular systems [4–11]. For example, in [12], there are more than 50 different interpretations of the "sustainable development" concept.

The implementation of sustainable development implies that certain monitoring of the studied system or phenomenon should be carried out. Monitoring is understood as a system of constant overseeing by the current of any phenomenon for the establishment of its compliance to the initial assumptions or desirable result. This phenomenon can occur in any sphere—in social relations, in nature, in the financial and economic sphere, etc. Within monitoring, there is assessment, control of the system, and the formation of management recommendations (management of its state) depending on the impact of particular factors.

Therefore, the formulation of the formalized concept of monitoring sustainable development, which could be concretized for particular cases, seems to be an urgent problem. One of the possible ways is to use a systems approach [13]. Let us make an attempt to formulate monitoring of the sustainable development concept of complex systems with the example of territories.

2. Problem statement

The systems approach involves the representation of the system S in the form of interrelated elements (infrastructures, key indicators, etc.). Territorial systems are complex ecological and socioeconomic systems, consisting of a large number of interacting elements. They are characterized by a huge variety and complexity of factors, elements of infrastructures, and relations between them [14]. The features of territorial systems can be attributed to [15] multidimensionality, the interconnectedness of components, stochastic nature of the behavior, multicriteriality, and diversity behavior of the elements.

Taking these features into account, the system S can be represented as a random vector $\mathbf{Y} = (Y_1, Y_2, \ldots, Y_m)$. Each component Y_i of this vector is a one-dimensional random variable characterizing the functioning of the corresponding element of the system (**Figure 1**).

The sustainable development concept has to reflect two components: "sustainability" and "development." Therefore, for its formulation, it is necessary to solve the following three tasks:

1. The development of an integral indicator which characterizes the effective functioning of the system.

2. Ensuring the sustainable functioning of the system.

3. The formation of a criterion which characterizes a sustainable development in the point of solving the first two tasks.

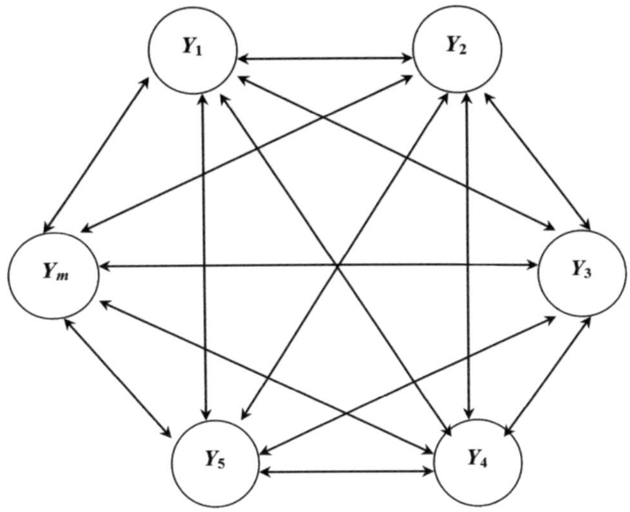

Figure 1.
A model of the system as a structure model.

3. Vector entropy model for the effective functioning of systems

Consider the problem of developing an integral indicator that characterizes the effective functioning of the system.

Multicriteriality of complex systems functioning, including territorial, and the diversity of their elements functioning, makes the development of universal formal indicators difficult which characterizes the effectiveness of systems as a whole.

It is known that entropy is a fundamental property in any systems with probabilistic behavior [16]. The concept of entropy is flexible and allows interpretation in terms of the branch of science, where it is applied. Therefore, entropy modeling is one of the promising lines of research of complex stochastic systems [17–20].

However, the frequent use of entropy for modeling of open systems, in contrast to thermodynamics, is insufficiently formalized and has generally qualitative and private character; there are no rather simple and adequate mathematical models that allow associating entropy with the actual characteristics of conditions of multidimensional systems. Common in these works is the use of Shannon's information entropy [21]. But, as it is noted in [15], the information entropy allows developing adequate entropy models only for particular problems.

However, in the same work [21], Shannon heuristically offered a formal analog of a concept of information entropy for the m-dimensional continuous random vector of Y with a probability density:

$$H(\mathbf{Y}) = -\int_{-\infty}^{+\infty} \dots \int_{-\infty}^{+\infty} p_{\mathrm{Y}}(y_1, y_2, \ \dots \ , y_m) \ln p_{\mathrm{Y}}(\mathbf{y}) dy_1 dy_2 \dots dy_m. \qquad (1)$$

This value Kolmogorov together with Gelfand and Yaglom was called subsequently differential entropy [22].

The differential entropy, being the functional given on the set of the probability density of a random vector of Y, represents a number. Therefore it cannot be an adequate mathematical model of a multidimensional system. However, the practical use of entropy (1) is complicated by the need to know the distribution law of a multidimensional random value of Y.

In [23] it was offered to use a differential entropy (further, an entropy) for modeling multidimensional stochastic systems. It is proven [23] that entropy in Eq. (1) can be represented as a sum of two components:

$$H(\mathbf{Y}) = H(\mathbf{Y})_V + H(\mathbf{Y})_R, \tag{2}$$

$$H(Y)_V = \sum_{i=1}^{m} H(Y_i) = \sum_{i=1}^{m} \ln \sigma_{Y_i} + \sum_{i=1}^{m} \kappa_i \text{—randomness entropy,}$$

$$H(Y)_R = \frac{1}{2}\sum_{k=2}^{m} \ln\left(1 - R^2_{Y_k/Y_1Y_2\ldots Y_{k-1}}\right)\text{—self-organization entropy,}$$

where $\sigma^2_{Y_i}$—dispersion, $\kappa_i = H(Y_i/\sigma_{Y_i})$—entropy indicator shows a type of random value distribution law Y_i, $i = 1,2,\ldots,m$; $R^2_{Y_k/Y_1Y_2\ldots Y_{k-1}}$ - coefficient of determination of regression dependencies of random vector $\mathbf{Y}$, $k = 2,3,\ldots, m$.

In particular, for multidimensional normally distributed random variable $\mathbf{Y}$.

$$H(\mathbf{Y})_V = \sum_{i=1}^{m} \ln \sigma_{Y_i} + m \ln \sqrt{2\pi e}, H(Y)_R = \frac{1}{2}\ln(|R|), \tag{3}$$

where $\mathbf{R}$ is the correlation matrix of random vector $\mathbf{Y}$.

The formula (2) does not always explain the behavior of the system. The addition of the component $H(\mathrm{Y})_V$ and $H(\mathrm{Y})_R$ in terms of systems analysis is incorrect since they characterize various regularities of the complex systems: $H(\mathrm{Y})_V$ is additivity, and $H(\mathrm{Y})_R$ is an integrity of the system.

The practical use of the relation (2) showed that there are situations when systems with different functional states have approximately the same general entropies of $H(\mathrm{Y})$, but the corresponding values of entropies of randomness $H(\mathrm{Y})_V$ and self-organization $H(\mathrm{Y})_R$ have significant differences. It schematically looks as follows. There are two the same systems $\mathrm{Y}^{(1)}$ and $\mathrm{Y}^{(2)}$ with different states. At the same time, $H(\mathbf{Y}^{(1)}) = 0$, $H(\mathbf{Y}^{(1)})_V = 1$, $H(\mathbf{Y}^{(1)})_R = -1$ and $H(\mathbf{Y}^{(2)}) = 0$, $H(\mathbf{Y}^{(2)})_V = 10$, $H(\mathbf{Y}^{(2)})_R = -10$.

Complex systems, including territorial ones, are open, and their entropy can both increase and decrease. Moreover, the directions of change in the entropies of randomness $H(\mathbf{Y})_V$ and self-organization $H(\mathbf{Y})_R$ of systems may be different. To build adequate models and investigate multidimensional stochastic systems, differential entropy should be considered not in scalar, but in vector form as two components—the entropies of randomness and self-organization as [15]:

$$h(\mathbf{Y}) = (h_V;\ h_R) = \left(H(\mathbf{Y})_V;\ H(\mathbf{Y})_R\right). \tag{4}$$

In specific situations, the direction and values of the entropy vector Eq. (4) should be set on the basis of the features of the studied system. In other words, complex systems should have a balance between the entropies of randomness and self-organization.

The complex systems are open. Influence of entropy on the evolution of open systems was investigated by many scientists. In their publications, it is noted that the change of open systems either leads to degradation or it is self-organization process as a result of which more complex structures appear. Prigogine [24] in 1955 formulated an extended version of the second law of thermodynamics. According to this law, the total change of entropy dS of an open system must be represented in the form of two parts. The reason of the first of them serves internal processes which are irreversible and by all means are followed by the transition of a part of the energy of ordered processes (kinetic energy of a moving body, energy of electric current, etc.) into the energy of the disordered processes and eventually in warmth. The second part is caused by the exchange of energy and substance between a system and a surrounding medium:

$$dS = dS_{in} + dS_{out}, \tag{5}$$

where dS is the total change in entropy of an open system, dS_{in} is the change in entropy during the processes occurring in a system, and dS_{out} is the change of entropy during the processes of exchange with the environment.

However, the question of the practical application of this theory for research of real systems has not been disclosed. Let us express the change of the total entropy of ΔH (Y) through the changes of entropies of randomness and self-organization:

$$\Delta H(\mathbf{Y}) = \Delta H(\mathbf{Y})_V + \ \Delta H(\mathbf{Y})_R \tag{6}$$

Let us try to give an interpretation of Eq. (5) according to Eq. (6). First, it is apparent that $dS \cong \Delta H(Y)$. The sign of the conditional equality "$\cong$" is used in view of the fact that in [24], change of a thermodynamic entropy of dS was considered.

Let us consider the influence on the entropy of processes of exchange with the environment. From the environment multidimensional open system takes or gives energy, which can be treated as a change of mean square deviations σ_{Y_i}. Besides the appearance of new properties, states can also occur from the outside, from the environment. Therefore, the change of distribution type, and therefore entropy indicators, is also due to the process of exchange of the system with the environment, that is, it is possible to consider that the change of entropy during processes of exchange with the environment represents a change of randomness entropy:

$$dS_{out} \cong \Delta H(Y)_V. \tag{7}$$

System elements in the process of functioning can strengthen or weaken the interaction between them due to the increase or decrease of the narrowness of correlation communication. Therefore, the change of entropy during the processes happening in a system is a change of self-organization entropy:

$$dS_{in} \cong \Delta H(Y)_R. \tag{8}$$

On the basis of Eqs. (7) and (8), it is possible to make the following hypothesis.

Hypothesis 1. The total change of entropy of an open system consists of the sum of two items. The first item characterizes the impact of the interaction of a system with the external environment and represents a change of randomness entropy. The second item characterizes the processes occurring within a system and represents a change of self-organization entropy.

Example 1. Entropy analysis of Yekaterinburg (regional center of Sverdlovsk region) development dynamics in 1992–2017.

The effective functioning of the megalopolis as a complex system according to the vector entropy model Eq. (4) consists in the simultaneous growth of diversity, opportunities for all elements of this system, and the presence of a close interrelation between these elements. This is manifested in the fact that with the development of a megalopolis, its randomness entropy should gradually increase, and the self-organization entropy will decrease.

The analysis will be performed according to the official data from the Russian Federal State Statistics Service (Rosstat) [24]. Of the many basic socioeconomic indicators of cities, we will form a system of signs that characterize all the main aspects of the city's infrastructure [15]:

1. Natural increase, decrease (−) per 1000 population.

2. The share of the working population in organizations, %.

3. Average monthly nominal accrued wages (in 2017 prices), thousand rubles.

4. The share of retirees registered with the social security authorities, %.

5. The total area of residential premises per one urban resident (at the end of the year), m^2.

6. Number of pupils in preschool educational organizations, thousand people.

7. The number of doctors per 1000 population, people.

8. The number of registered crimes per thousand people.

9. The volume of work performed under construction contracts (in 2017 prices), ths. rub. for one person.

10. Retail trade turnover (in 2017 prices), ths. rub. for 1 person.

11. Investments in fixed assets (in 2017 prices), ths. rub. for one person.

When calculating the entropy, the estimates were performed for periods of 13 years. This period turned out to be optimal, on the one hand from the statistical smoothing point of view and on the other hand because it takes into account the dynamics of entropy change. Entropy was estimated in the vector form Eq. (4). Accounting for inflation was carried out by recalculation in 2017 prices based on consumer price indices; the different populations of cities were taken into account by the transition to relative indicators per inhabitant. Since the sample was quite small, the deviations of the empirical distributions of the considered features from the normal distribution are practically impossible to establish. Therefore, when calculating the entropies of randomness and self-organization, we use Eq. (3).

Figure 2 shows the graphs of changes in the entropies of randomness and self-organization in Yekaterinburg. **Figure 3** shows the entropy dynamics.

Analysis of graphs in **Figures 2** and **3** allows us to make the following conclusions:

1. The period of stabilization of operation (until 2008). The randomness entropy h_V increases, and the self-organization entropy h_R decreases.

2. The global financial crisis of 2008–2009. Short-term sharp change in the direction of vector entropy to the opposite.

3. The period of economic recovery after the financial crisis (2009–2011), followed by a decrease in the growth rate of gross domestic product

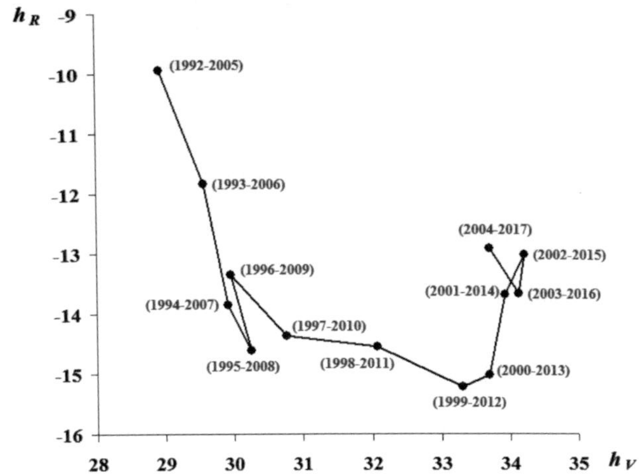

Figure 2.
The change in self-organization and randomness entropies in Yekaterinburg.

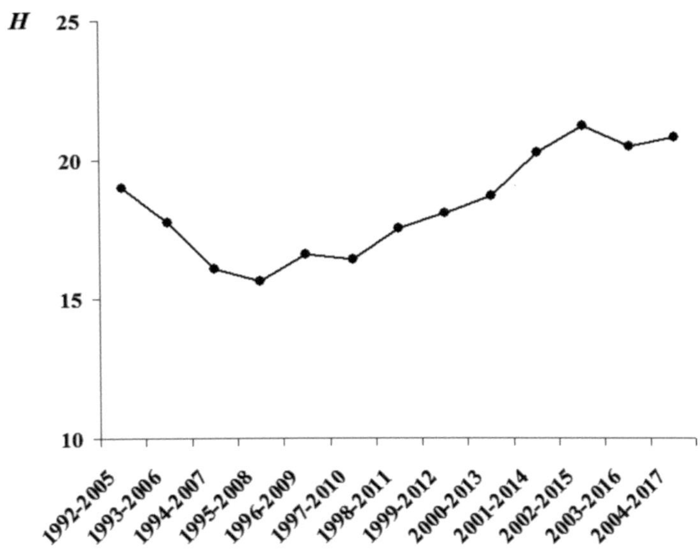

Figure 3.
Entropy dynamics in Yekaterinburg.

(2012–2013). The direction of vector entropy gradually changes (at first h_V increases, and h_R decreases, but gradually this trend fades).

4. Announcement of sanctions (2014). A steep increase in h_R.

5. Functioning in terms of sanctions (since 2014). There is a practical lack of vector entropy dynamics, with the value of self-organization entropy h_R fixed at the level of 2006–2007, and randomness entropy h_V at the level of 2013.

6. The total entropy $H(\mathbf{Y})$ in the period under consideration has changed slightly.

Example 2. The modeling of a system that characterizes the safety of the production.

Seventeen coal mining enterprises were investigated [23]. On the basis of two generalized factors (Y_1 is the factor characterizing the organization of safe production; Y_2 is the factor reflecting the professionalism of the staff), all enterprises were divided into two groups: (1) enterprises with a low level of injury; (2) enterprises with a high level of injury. For the first and second groups of mines, respectively, we have.

$$\left(h_V^{(1)};h_R^{(1)}\right) = (2,42;\ -0,31), H\left(Y^{(1)}\right) = 2,11,$$

$$\left(h_V^{(2)};h_R^{(2)}\right) = (3,74;\ -0,70), H\left(Y^{(2)}\right) = 3,04.$$

In this case, the direction of the entropy change vector will differ from Example 1: at the enterprises with a high level of injury, the randomness entropy needs to be reduced, and the self-organization entropy needs to be increased.

For example, this can be accomplished as follows: to bring the state of the second group of mines to the state of the first group, it is necessary to reduce the dispersion of the factor characterizing the organization of safe production and reduce interrelation with the factor reflecting the professionalism of the staff. This means a more specific and accurate organization of production. The organization of safe production should not depend on the degree of professionalism and competence of staff strongly.

Example 3. We investigate the possibilities of entropy modeling on the example of the population analysis in terms of prevention of chronic noninfectious diseases (CNID) by biological risk factors [23]. For carrying out the analysis of change of population entropy depending on the health status, two equal age groups were formed: 18–26 years and 27–35 years. Four risk factors were identified: "total cholesterol," "systolic blood pressure," "body mass index," and "glucose level." The results of the analysis are given in **Table 1**.

As the health of the population deteriorates, the total population entropy and the randomness entropy increase. This can be explained by the fact that the additional damaging influence of CNID, in general, is added to the pathological influence of risk factors on a human body separately and on all population.

Conversely, the self-organization entropy as the deterioration of the health status of the population decreases which corresponds to the strengthening of the narrowness of the interrelations between subsystems. This can be explained by the fact that the development of diseases in the organism happens in many respects and it is interdependent. On the other hand, at the development of diseases, some

Age (years)	Health status	Randomness entropy $H(Y)_V$	Self-organization entropy $H(Y)_R$	Total entropy $H(Y)$
18–26	Healthy	5.500	−0.514	4.986
	Apparently healthy	7.131	−0.578	6.553
	Patient	7.847	−0.696	7.151
27–35	Healthy	5.731	−0.299	5.432
	Apparently healthy	8.376	−0.542	7.834
	Patient	8.720	−0.781	7.939

Table 1.
Entropy levels in different groups of people.

subsystems can adapt to others, compensating shortcomings their functioning (substitution effect).

4. Multidimensional risk model of complex systems

Consider the task of ensuring the system sustainable functioning. Here you first need to concretize the "sustainability" concept. Typically the stability of the system functioning is interpreted in terms of its safety. Security issues are resolved with the help of risk analysis [25]. Some authors note that growth rates of damage considerably exceed growth rates of the economy [26, 27]. This can be explained with a constant increase of risk in the conditions of a scientific and technical revolution and the forced development of a technosphere [28]. Therefore, we will assume that the functioning stability of the territorial system is intimately connected with risk; the lower the risk level, the more stable the system state. Thus, the diagnosis of system sustainability can be made on the basis of monitoring its risk. This requires adequate risk models.

Let S be some multidimensional stochastic system. Let us consider an adequate representation of this system as a random vector $\mathbf{X} = (X_1, X_2, \ldots, X_l)$ with a certain probability density $p_{\mathbf{X}}(\mathbf{x})$. The development of a complex system and an increase in the efficiency of its functioning are an inevitable cause of increasing risks. Therefore, it is necessary to assess the risks of such systems. Consider the risk model of multidimensional stochastic systems proposed in [29].

Instead of the generally conventional selection of concrete dangerous situations, we will define the geometric area D of adverse outcomes. Formally this area can look arbitrarily depending on a specific objective.

The concept of dangerous states as larger and improbable deviations of a conception of dangerous conditions as large and improbable deviations of random variables from some best provision $\mathbf{\Theta}$ is mostly distributed. In this case, D represents an external area of an m-axis ellipsoid.

Setting the function of consequences from dangerous situations (risk function) in the form of $g(x)$, we will receive a model for the quantitative assessment of risk [30]:

$$r(X) = \int\int_{R^m} \ldots \int g(x) p_X(x) dx. \tag{9}$$

If in Eq. (9) to accept $g(x) = 1\ \forall x \in D$ and $g(x) = 0\ \forall x \notin D$, that $r(X) = P(X \in D)$, that is, the risk is estimated as a probability of an unfavorable outcome.

If at an early stage of system analysis is difficult to describe enough precisely the $g(\mathbf{x})$ function, then Eq. (2) becomes an assessment of $P(D)$ and is a convenient initial approximation of risk model.

To define a function $g(\mathbf{x})$ requires a quantitative assessment of consequences for the studied system depending on values of risk factors. It demands to carry out separate research. Let us note that values of the function $g(\mathbf{x})$ are given in the nominal units. But they are usually quite simply interpreted in the respective subject area. The essence of the function $g(\mathbf{x})$ is as follows. It accepts the least nonnegative (e.g., zero) value in a point of $\mathbf{\Theta}$ or in its neighborhood of $U(\mathbf{\Theta})$. Further in each direction during removal from $U(\mathbf{\Theta})$ the $g(\mathbf{x})$ function has to increase monotonously. For scaling on each risk factor, we will set some limit values, at which consequences become dangerous (or irreversible). Let us set values $g(\mathbf{x})$ at each such point equal to some value, for example 1. For convenience, it is desirable to impose a number of restrictions on the function $g(\mathbf{x})$: convexity, continuity, etc.

In [30] the variant of the task of the $g(\mathbf{x})$ function in the form of a paraboloid is given. By way of illustration in **Figure 4**, the example of the risk function for a case $m = 2$ is shown. The ellipse describing the area $\overline{D}$ of admissible values of risk factors and lying on the Ox_1x_2 ($r = 0$) plane is shown by black color. The paraboloid above the plane represents possible values of risk $r(\mathrm{X})$. White points on the plane are values of risk factors; to them there correspond points on paraboloid surface which set risk values; the image of the border of an ellipse $\overline{D}$ is shown in the form of the black line. All corresponding couples of points (values of risk factors and risk values) are connected among themselves by vertical dashed lines.

In the problems of risk monitoring, along with risk assessment, $r(X)$ on all risk factors of X_1, X_2 ..., X_m of the multidimensional system is expedient to estimate the contribution of each factor to total risk. We introduce a random vector $X_k^- = (X_1, \ldots, X_{k-1}, X_{k+1}, \ldots, X_m)$. Then the absolute change of risk of the multidimensional system due to the addition of factor X_k is equal:

$$\Delta r(X_k) = r(X) - r\left(X_k^-\right). \tag{10}$$

Dividing $\Delta r(X_k)$ of the risk $r\left(X_k^-\right)$, we will receive the relative change of risk of the multidimensional system by the addition of factor X_k:

$$\delta r(X_k) = \Delta r(X_k) / r\left(X_k^-\right). \tag{11}$$

Let us note that along with a contribution to the common risk of one factor, Eqs. (10) and (11) allow us to estimate influence and groups of factors.

Monitoring risk on the basis of the model in Eqs. (9)–(11) consists of serial estimation in time of the actual values of $r(\mathbf{X})$, $\Delta r(X_k)$, $\delta r(X_k)$, $j = 1, 2, \ldots, m$, and also dynamics of their change.

Let us consider the most common case when $\mathbf{X}$ has a joint normal distribution with a probability density:

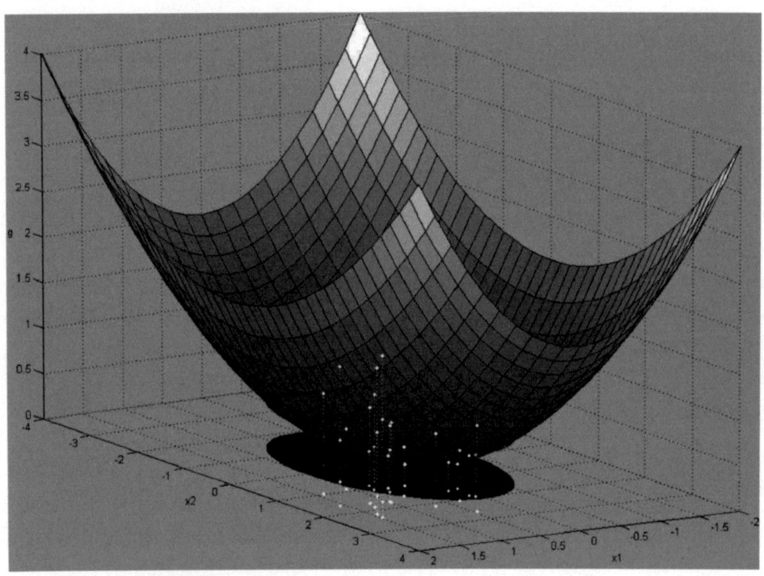

Figure 4.
An example of a two-dimensional risk functions.

$$p_X(x) = \frac{1}{\sqrt{(2\pi)^m |\Sigma|}} \exp\left\{-\frac{1}{2}(x-a)^T \Sigma^{-1}(x-a)\right\},$$

where $a = (a_1, a_2, .., a_m)^T$—a vector of expectations, $\Sigma = \{\sigma_{ij}\}_{m\times m}$—a covariance matrix, $\sigma_{ii} = \sigma_i^2$—dispersion of factor X_i.

The use of a Gaussian random vector is based on the central limit theorem [31]. As approbation on a number of examples has shown, such idealization is not so critical, and if there are any bases to consider that density of probabilities is a component of the vector of **X** having more extended tails, then this can be corrected by setting the $g(\mathbf{x})$ function accordingly.

Example 4. Let us consider a two-dimensional Gaussian random vector which components have a zero average and single dispersion. In **Figure 5** the example of realization of such accidental vector is shown for: (a) $\rho = 0,4$; (b) $\rho = 0,9$, where ρ is a coefficient of correlation between *X1* and *X2*.

From **Figure 5** we see that the probability of large deviations of the random vector from the origin increases with the increase in the closeness of the correlation.

Example 5. Let us estimate the probability of P(D) for a Gaussian random vector of X, with the different narrowness of correlation communication $D_e(X) = 1 - |R_X|^{1/m}$ [32], where $|R_X|$ is a continuant of a complete correlation matrix ($D_e(X) = 0$ is independence of components, and $D_e(X) = 1$ is a rigorous linear relation). Let us consider the following cases: $D_e(X) = 0$, $D_e(X) = 0,5$, $D_e(X) = 1$. The results of the calculation of $P(D)$ are given in **Figure 3**. For descriptive reasons we will accept $A_1 = A_2 = ... = A_m = A$, $A_j = b_j/\sigma_j$.

The analysis of schedules in **Figure 6** indicates the following.

The increase in the probability of an unfavorable outcome is influenced by both the increase in the dimension of m and an increase in the narrowness of correlation communication between the components of a random vector of X. Let us note that even the average narrowness of correlation ($D_e(X) = 0,5$) leads to a significant increase in the probability of an unfavorable outcome. The effect increases with the increase in values A_j that correspond to less probable, but more dangerous, adverse outcomes. Therefore, risk modeling should take into account both a multidimensionality factor and narrowness of correlations.

Example 6. Approbation of risk model of a multidimensional stochastic system for monitoring of risk of Sverdlovsk region in 1999–2017.

Let us execute monitoring of Sverdlovsk region on the dynamics of macroeconomic risk factors, taken as an interval of 9 years. Risk factors and their threshold levels are given in **Table 2**.

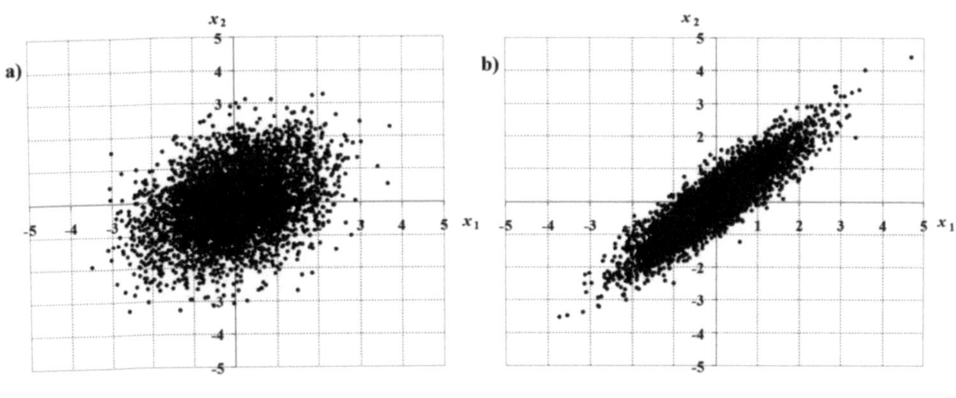

Figure 5.
Realization of a standard normal random vector.

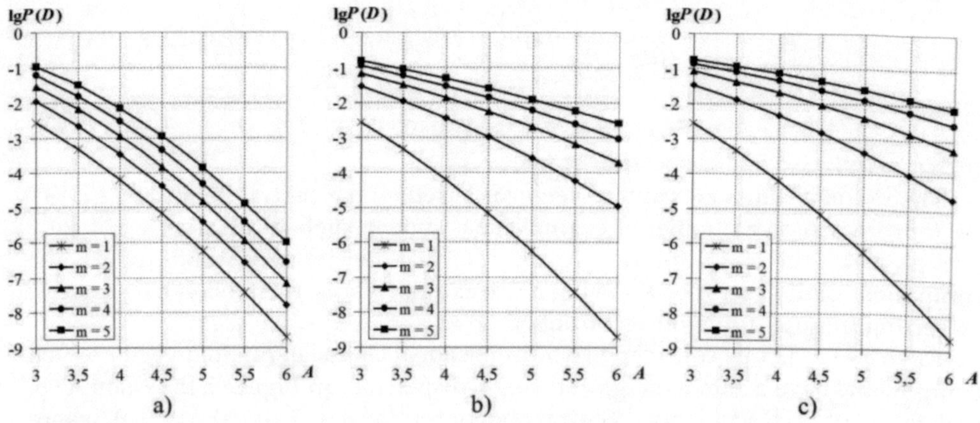

Figure 6.
Dependences of lgP(D) on threshold level A: (a) $D_e(X)$ = 0; (b) $D_e(X)$ = 0.5; (c) $D_e(X)$ = 1. Designations: Row 1 (m = 1), row 2 (m = 2), row 3 (m = 3), row 4 (m = 4), row 5 (m = 5).

Risk factors	Threshold levels K_j
X_1—real income movement, in % to the previous year	79.93
X_2—the ratio of the average size of pension to subsistence minimum of pensioners	0.66
X_3—morbidity on 1000 people of the population	960
X_4—mortality from external causes, number of the dead on 100,000 people of the population	322.1
X_5—wear of fixed assets on the end of the year, %	71.33
X_6—the volume of budget revenues per capita, in the prices of 2017, thousand rubles	21.75
X_7—quantum index of gross regional product, % to the previous year	88.4
X_8—unemployment rate, in %	18

Table 2.
Macroeconomic risk factors of the region.

We consider that random vector **X** has a joint normal distribution.

In **Figures** 7 and 8, results of the calculation of the probability of unfavorable outcome *P(D)* and risk $r(\mathbf{X})$ for the threshold levels of risk factors K are shown.

Analysis of the results of monitoring of multidimensional risk in the Sverdlovsk region showed the following:

1. During the initial period, the greatest socioeconomic instability (the highest risk values) was observed. Then gradually the dynamics of sustainability in the region have increased (decreased risks).

2. After the sanctions were imposed, the risk began to increase. The lower rate of growth of $r(\mathbf{X})$ than $P(D)$ indicates that the probability of occurrence of very dangerous situations has been growing slightly since 2014.

3. In the Sverdlovsk region until 2010, the main contribution to regional instability was made by factor X_6, then X_2 became such a factor, and since 2015, the main contribution to instability was made by factor X_7.

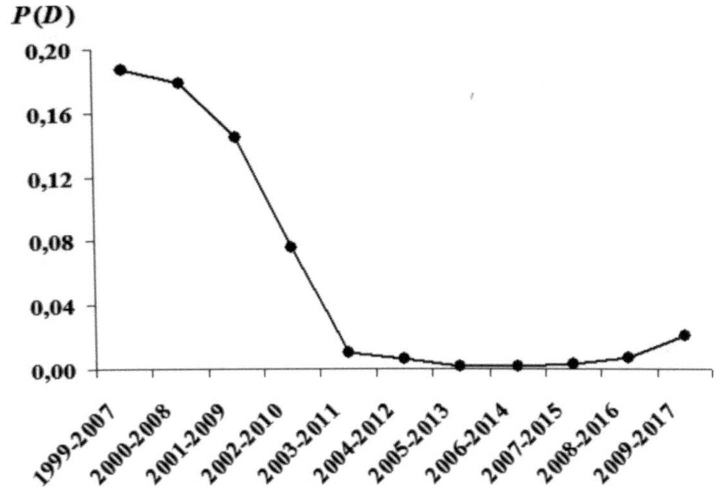

Figure 7.
Estimates of P(D) in the Sverdlovsk region.

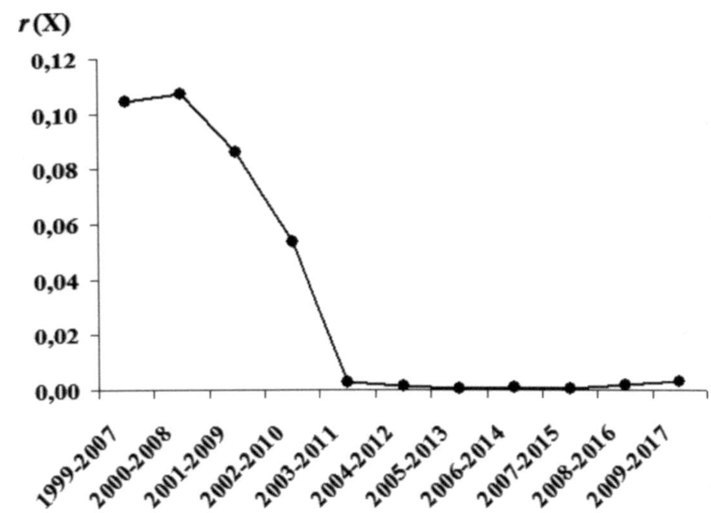

Figure 8.
Estimates of $r(\mathbf{X})$ *in the Sverdlovsk region.*

5. Formation of the sustainable development concept

Consideration of examples shows that “development” and “sustainability” characterize various aspects of the complex systems operation. And for ensuring sustainable development, they need to be taken into account together.

Hypothesis 2. We will understand dynamics consisting of available trends of the balanced change of a vector entropy while maintaining an acceptable level of risks as sustainable development of the complex system.

For this purpose, we combine the vector entropy model and the risk model of a multidimensional stochastic system. Moreover, it is necessary to consider elements of the system (components of the random vector $\mathbf{Z}$), both as risk factors X_i and as indicators Y_j, characterizing the functioning of the system, that is,

$$\mathbf{Z} = \mathbf{X} \cup \mathbf{Y} = (Z_1, Z_2, \ldots, Z_n), \ \max(l, m) \le n \le l + m.$$

The case $n < l + m$ appears, when $X \cap Y \neq \varnothing$.

Within the framework of the proposed concept, along with the tasks of monitoring complex systems discussed above, it is possible to solve management problems (development of control recommendations).

The idea of vector entropy control is the transfer of the vector $\mathbf{h}(\mathbf{Z})$ from the state $h(Z^0) = (h_V^0;\ h_R^0)$ in the state $h(Z^*) = (h_V{}^*;\ h_R{}^*)$, which corresponds to the effective functioning of the stochastic system.

For a Gaussian system, the vector entropy control consists in directing the entropy from some initial point $(h_V^0;\ h_R^0) = (H(Z^0)_V;\ H(Z^0)_R)$ with the covariance matrix Σ_0 to the final point $(h_V{}^*;\ h_R{}^*)$ with a minimal change of the covariance matrix$\Sigma^0 = \left\{\sigma_{ij}^0\right\}$ and the expectations vector a^0 and acceptable risk (**Figure 9**).

The problem of the vector entropy control of the Gaussian system to ensure sustainable development will take the form:

$$\begin{cases} G(\Sigma) = \sum_{i=1}^{n}\sum_{j=1}^{n}\left(\sigma_{ij} - \sigma_{ij}^0\right)^2 + \sum_{i=1}^{j}\left(a_i - a_i^0\right)^2 \to \min_{a,\Sigma}, \\ H(Y)_V = A, \\ H(Y)_R = B, \\ r(X) \le r, \\ \Sigma \in G_\Sigma,\ \ a \in H_a, \\ \sigma_{ij}^2 < \ \sigma_{ii}\sigma_{jj},\ \ \sigma_{ij} = \sigma_{ji},\ \ \sigma_{ii} > 0\ \ \forall\ 1 \le i, j \le n, \\ \Sigma > 0, \end{cases} \tag{12}$$

where $A = h_V{}^*$, $B = h_R{}^*$, **a**—the expectations vector of the components X_i, $i = 1, 2, \ldots, l$.

The last constraint in Eq. (12) means positive definiteness of the matrix Σ. Note that the performance criterion in Eq. (12) may be different, depending upon the characteristics of a particular system S.

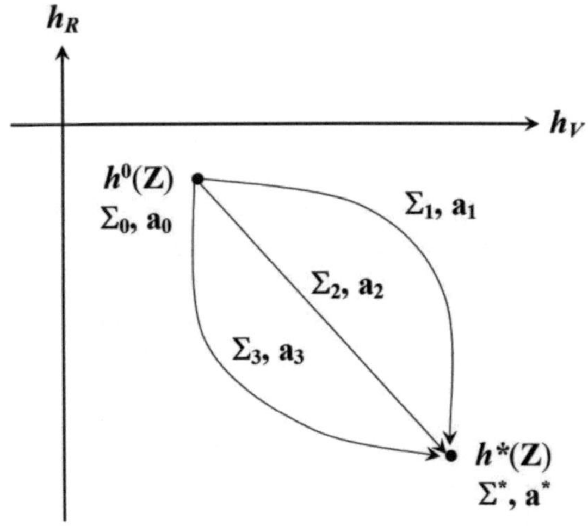

Figure 9.
An illustration of vector entropy management to ensure sustainable development of the system.

6. Discussion of results

Thus, on the basis of the use of two original models—a vector entropy and multidimensional risk—it was succeeded to formalize the new concept of sustainable development of complex systems. Both models are successfully approved on real data.

This concept can be implemented by means of monitoring of the studied system. As observed parameters efficiency factors of the functioning of a system and its risk factors are used. The direction of development is given by an entropy vector, and stability is provided due to an acceptable risk level.

Management recommendations are formed in the form of the solution to an extreme problem Eq. (12). This problem is solved by methods of penalty functions. Currently, the work is at a stage of practical approbation of monitoring of sustainable development of Sverdlovsk region.

7. Conclusion

1. The probability-entropy concept of sustainable development of complex stochastic systems is formulated. It is based on vector entropy and multidimensional risk models.

2. According to the formulated concept, the sustainable development of a complex system will be understood as the dynamics consisting of the tendency of a balanced change in vector entropy while maintaining an acceptable level of risk.

3. The proposed concept of sustainable development has been tried out in application to territorial systems.

Acknowledgements

The present study was carried out by financial support of the Russian Fund of Fundamental Research grant № 17-01-00315 “Development of models and methods of monitoring, forecasting, and control optimization of multidimensional regional socioeconomic systems based on entropy and minimax approaches.”

Author details

Alexander Nikolaevich Tyrsin[1,2], Al'fiya Adgamovna Surina[3*]
and Alexey Sergeyevich Antonov[1]

1 Ural Federal University named after the first President of Russia B.N. Yeltsin, Yekaterinburg, Russia

2 Science and Engineering Center "Reliability and Resource of Large Systems and Machines", Ural Branch, Russian Academy of Sciences, Yekaterinburg, Russia

3 South Ural State University (National Research University), Chelyabinsk, Russia

*Address all correspondence to: dallila87@mail.ru

References

[1] Our Common Future. Report of 1983 World Commission on Environment and Development (WCED) Доклад Всемирной комиссии по вопросам окружающей среды и развития [Internet]/UN General Assembly. Available from: https://www.un.org/ru/ga/pdf/brundtland.pdf [Accessed: 01-07-2019]

[2] Begun TV. Sustainable development: Definition, concept, and factors in the context of monotowns. In: Economics, Management, Finance: Proceedings II Int. Scientific Conf. Perm: Mercury; 2012. pp. 158-163. Available from: http://moluch.ru/conf/econ/archive/57/3117/

[3] Peregudov FI, Tarasenko FP. Introduction to System Analysis. Moscow: Higher School Publ; 1989. p. 367. (in Russian)

[4] Barbier E. The concept of sustainable economic development. Environmental Conservation. 1987;**14**(2):101-110

[5] Costanza R, Daly HE. Natural capital and sustainable development. Conservation Biology. 1992;**6**(1):37-46

[6] Loguntsev EN. The Concept of Sustainable Development from the Standpoint of an Interdisciplinary Approach. City Government. № 11. 2000. pp. 24-32

[7] Granberg AG, Danilov-Danilyan VI. Strategy and problems of sustainable development of Russia in the XXI century. M: Economy; 2002

[8] Dyllick T, Hockerts K. Beyond the business case for corporate sustainability. Business Strategy and the Environment. 2002;**11**:130-141

[9] Williams CC, Millington AC. The diverse and contested meaning of sustainable development. The Geographical Journal. 2004;**170**(2): 99-104

[10] Ursul AD. Conceptual problems of sustainable development. Bulletin of RAS. Use and protection of natural resources in Russia. 2005;**1**:30-38

[11] Sachs JD. The Age of Sustainable Development. New York: Columbia University Press; 2015. p. 544

[12] Starikova EA. Modern approaches to the interpretation of the concept of sustainable development. Bulletin of the Peoples' Friendship University of Russia. Series: Economy. 2017;**25**(1):7-17

[13] Van Gigch JP. Applied General Systems Theory. 2nd ed. New York: Harper & Row Publishers; 1978

[14] O'Sullivan A. Urban Economics. 8th ed. Irwin: McGraw-Hill; 2012

[15] Tyrsin AN, Gevorgyan GG. Entropy modeling of sustainable development of megacities. IOP Conference Series: Earth and Environmental Science. 2017; **72**(9):012010. DOI: 10.1088/1755-1315/72/1/012010

[16] Nicolis G, Prigogine I. Self-Organization in Nonequilibrium Systems: From Dissipative Structures to Order through Fluctuations. New York: Wiley; 1977

[17] Wilson AG. Entropy in Urban and Regional Modelling. London: Pion; 1970

[18] Webber MJ. Information Theory and Urban Special Structures. London: Croom Helm; 1979

[19] Haken H. Information and Self-Organization: A Macroscopic Approach to Complex Systems. Berlin: Springer; 2000

[20] Ellis RS. Entropy, Large Deviations, and Statistical Mechanics. Berlin: Springer; 2006

[21] Shannon CE. A mathematical theory of communication. The Bell System Technical Journal. 1948;**27**:379-423, 623-656

[22] Gelfand IM, Kolmogorov AN, Yaglom AM. The Amount of Information and Entropy for Continuous Distributions. In the book: Works of the III All-Union Mathematical Congress. Vol. 3. Moscow: Academy of sciences of the USSR. 1958. pp. 300-320

[23] Tyrsin AN. Entropy Modeling of Multidimensional Stochastic Systems. Voronezh: Scientific book; 2016

[24] Prigogine I. Introduction to Thermodynamics of Irreversible Processes. New York: Wiley; 1955

[25] Mun J. Modeling Risk. 2nd ed. Wiley; 2010

[26] Vorobiev YL, Malinetsky GG, Makhutov NA. Risk management and sustainable development: The human dimension. News of universities. Journal of Applied Nonlinear Dynamics. 2000; **8**(6):12-26

[27] Gore A. Earth in the Balance. Forging a New Common Purpose. New York: Earthscan Publications Ltd., Copyright—Senator Al Gore; 1992

[28] Porfiryev BN. Emergency management. Results of science and technology. Security Issues: Emergencies. T. 1. (M: VINITI). 1991

[29] Tyrsin AN, Surina AA. Modeling of risk in multidimensional stochastic systems. Tomsk State University Journal of Control and Computer Science. 2017; **2**(39):65-72. DOI: 10.17223/19988605/39/9

[30] Tyrsin AN, Surina AA. Monitoring of risk of multidimensional stochastic system as tools for a research of sustainable development of regions. IOP Conference Series: Earth and Environmental Science. 2018;**177**: 012005. DOI: 10.1088/1755-1315/177/1/012005

[31] Gnedenko BV. The Theory of Probability. Moscow: Mir Publisher; 1978

[32] Pena D, Rodriguez J. Descriptive measures of multivariate scatter and linear dependence. Journal of Multivariate Analysis. 2003;**85**:361-374

Section 3

Methodologies about Sustainability Assessment

Chapter 4

Attributional and Consequential Life Cycle Assessment

Tomas Ekvall

Abstract

An attributional life cycle assessment (ALCA) estimates what share of the global environmental burdens belongs to a product. A consequential LCA (CLCA) gives an estimate of how the global environmental burdens are affected by the production and use of the product. The distinction arose to resolve debates on what input data to use in an LCA and how to deal with allocation problems. An ALCA is based on average data, and allocation is performed by partitioning environmental burdens of a process between the life cycles served by this process. A CLCA ideally uses marginal data in many parts of the life cycle and avoids allocation through system expansion. This chapter aims to discuss and clarify the key concepts. It also discusses pros and cons of different methodological options, based on criteria derived from the starting point that environmental systems analysis should contribute to reducing the negative environmental impacts of humankind or at least reduce the impacts per functional unit: the method should be feasible and generate results that are accurate, comprehensible, inspiring, and robust. The CLCA is more accurate, but ALCA has other advantages. The decision to make an ALCA or a CLCA should ideally be taken by the LCA practitioner after discussions with the client and possibly with other stakeholders and colleagues.

Keywords: life cycle inventory analysis, methodology, attributional LCA, consequential LCA, allocation, marginal data, electricity

1. Introduction

Life cycle assessment (LCA) is the quantification of potential environmental impacts and the resource use throughout a product's life cycle: from raw material acquisition, via production and use phases, to waste management [1]. It has been frequently applied by consultants, researchers, industry, and authorities for the past 30 years. It has proven useful for gaining knowledge on the life cycle, for communication of environmental information, and for various kinds of decision-making.

Meanwhile, it was clear almost from the start that results from different LCAs can contradict each other. This is still true, despite many attempts to harmonize, standardize, and regulate LCA. From history, we learn that it is not realistic to expect LCA to deliver a unique and objective result. It should not be regarded as a single unique method; it is more fruitful to consider it a family of methods.

Attributional LCA (ALCA) and consequential LCA (CLCA) are important groups within this family of methods. The choice between ALCA and CLCA guides other methodological decisions in the LCA, such as the choice of input data and the modeling of processes with multiple products. However, within ALCA and CLCA, there are still many decisions to be made—many versions or members within each group in the LCA family.

The purpose of this chapter is to discuss and clarify key concepts in relation to ALCA and CLCA and to guide the reader through the necessary and subjective methodological choices. The example used often relates to the supply of electricity in the life cycle, because much of the methodological debate has been on how to model electricity. The chapter is still relevant to all kinds of LCA, because energy supply is part of virtually all LCAs and because most of the discussion is valid also for modeling other parts of the life cycle. Furthermore, the chapter is relevant to other, similar types of quantitative environmental and sustainability assessments—for example, carbon footprint, which essentially is an LCA except that it is limited to emissions of greenhouse gases [2].

To structure the discussion on the pros and cons of different methodological choices, I start by establishing a set of criteria for what an LCA, or a quantitative environmental systems analysis in general, should be and do (Section 2). The ALCA and CLCA approaches are outlined in Section 3, and their implications for the choice of data and allocation problems are discussed in some detail in Sections 4 and 5, respectively. Section 6 includes an assessment of the two approaches based on previous discussions. The chapter concludes with a few recommendations for the LCA practitioner.

The LCA methodology is diverse, and the interpretation of the key concepts also varies between researchers. This chapter presents my view on the matter, which is subjective but based on knowledge gained from more than three decades of research in LCA and energy systems analysis. I present my arguments for this view but leave it to you, the reader, to accept my view or to choose another perspective.

2. Criteria for methods in environmental systems analysis

Environmental systems analysis is different from traditional science in that the aim is not just to systematically gather knowledge; it has the specific aim to gather and communicate knowledge that results in actions that reduce the negative environmental impacts of human activity in total or at least per functional unit, that is, per unit of utility that the studied system generates. The more a method for environmental assessments can be expected to contribute to this purpose, the better it is.

For a method to benefit the environment, it must be possible to apply. The results need to be reasonably accurate, possible to communicate, and perceived as relevant by decision-makers (**Figure 1**). Furthermore, the method should be resistant against abuse. Each of these criteria is briefly discussed below.

Different methods meet the criteria to varying degrees, but no method is ideal from all aspects. There will always be a trade-off between, for example, feasibility and accuracy. Hence, the set of criteria is not sufficient as a tool for objective selection of the best methods; however, it can be used for structured discussions on the pros and cons of available methods.

2.1 Feasible

To have any effect, the method must be used. The more often it is used, the more results it will generate. How often environmental assessments are made depends on how useful the results are (see Section 2.2–2.4). But it also depends on how easy the methods are to apply and how expensive the studies become. This in turn depends on how complex the methods are and on the extent to which the data and

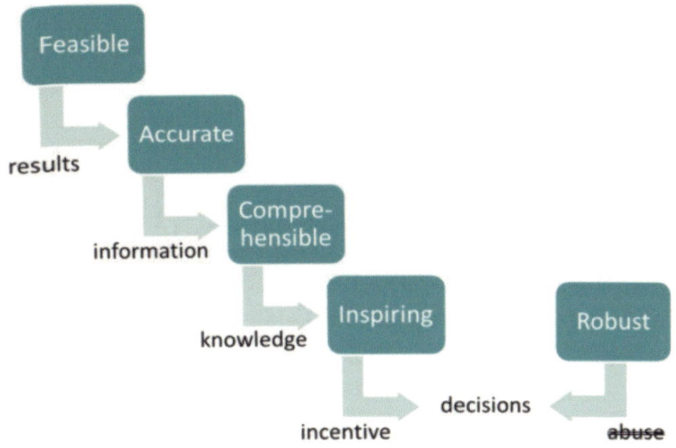

Figure 1.
Criteria for assessing methods for environmental assessment and their role in shaping decisions that are good for the environment (based on Ekvall et al. [3]).

models needed are available. The method becomes more cost-efficient and can potentially have a greater impact if the results and conclusions it generates can be generalized and reused in multiple decision situations. Hence, the method should ideally be easy and cheap to apply and generate results and conclusions that can be generalized.

2.2 Accurate

An environmental assessment is sometimes designed to guide a specific decision. To have a positive effect, the results must guide such decisions in the right direction more often than not. The greater the chance that results will point in the right direction, the better. Hence, the method should ideally generate results that are as comprehensive, accurate, and precise as possible.

2.3 Comprehensible

Besides guiding specific decisions, an environmental assessment can contribute to increasing the knowledge of experts and decision-makers. If accurate, such knowledge not only contributes to deliberate immediate actions but can also have a positive impact on future decisions. To educate decision-makers and other stakeholders, the environmental assessment must be transparent and possible to understand. Decision-makers receive a large amount of information and have limited capacity for information processing. For this reason, the method and the results it generates should be easy to communicate and understand. Communication is easier when the concepts used in the method are clear and intuitively easy to understand. Communication is more challenging when the study is very comprehensive or conceptually complex. Hence, the method should ideally result in studies that are transparent, have a simple structure, and use intuitively clear concepts.

2.4 Inspiring

In order for environmental assessment to have a positive effect, the information and knowledge they generate must result in actions. Decision-makers often have conflicting goals, and decisions are often not rational in the sense that they are based on documented facts only. To convince and inspire decision-makers, the study should be perceived as relevant, legitimate, and credible and the recommendations clear. A study can be perceived as more relevant if it focuses on things that the decision-makers can influence and/or have a clear connection to. Legitimacy increases if the study is perceived as impartial and fair. Credibility can be obtained, for example, through sensitivity analyses. The conclusions and recommendations are clear when the uncertainties are not too great.

Relevance and legitimacy are highly subjective. They both increase if the design of the study accounts for the need for knowledge as perceived by the decision-makers. This means that the choice of methods should ideally be adapted to the situation and may vary depending on the decision-makers involved.

2.5 Robust

Robustness here means that the method gives roughly the same results regardless of who applies it. This makes the method more difficult to abuse, that is, to apply in environmental assessments with the purpose to stop or delay decisions with positive consequences for the environment or to defend decisions with poor consequences. The method becomes more robust if it does not require the user to make assumptions or subjective choices that greatly affect the results. It is also more robust if there are detailed guidelines for how the method is to be applied and/or an established good practice for the application.

3. Attributional and consequential LCA respond to different questions

As clear from the Introduction, we can distinguish between ALCA and CLCA. The distinction between two types of LCA was suggested in the beginning of the 1990s [4, 5]. It was established toward the end of the decade [6] to resolve debates on what type of input data to use in LCAs (cf. Section 4) and on how to deal with the allocation problems that occur when, for example, a process produces more than one type of product (Section 5). Various names were used on the two types of LCA [7], but the terms attributional/consequential have been used since 2001 [8].

Several different definitions of attributional and consequential LCA have been suggested [9, 10]. I prefer the definitions of Finnveden et al., in what is probably the most cited scientific paper on LCA [11]:

- Attributional LCA: LCA aiming to describe the environmentally relevant physical flows to and from a life cycle and its subsystems

- Consequential LCA: LCA aiming to describe how environmentally relevant flows will change in response to possible decisions

These definitions clearly connect ALCA/CLCA not only to methodological choices but also to the goal of the study, because they respond to different questions (**Figure 2**). An ALCA gives an estimate of how much of the global environmental impact belongs to the product studied. A CLCA gives an estimate of how the global environmental impact is affected by the product being produced and used.

Note that the latter can include both increases and reductions in the environmental impact. It is not unusual that an increase in the production of a product leads to increases in emissions as well as to environmental benefits. The production of district heating in a combined heat-and-power (CHP) plant in Sweden, for example, generates emissions from the CHP plant but reduces emissions in other parts of the electricity system, when electricity from the power plant replaces other electricity production.

There are thus two types of LCAs, carbon footprints, etc.:

- Attributional assessments, which give an estimate of what part of the global environmental burdens belongs to the study object

- Consequential assessments, which give an estimate of how the production and use of the study object affect the global environmental burdens

The choice between ALCA and CLCA affects system boundaries. In the example of district heating from a CHP plant, a CLCA includes both the emissions from the CHP plant and the reduction in emissions from the electricity production displaced by electricity from the CHP plant. In general, when a production process delivers more than one type of products, the CLCA should take into account how the process is affected by a change in the of the product investigated. If it affects the production of other products from the process, the system should be expanded to include the effect of that change.

A more advanced CLCA can also include other types of consequences. An increased use of a material in the studied system can, for example, lead to less material being used in other systems. This reduction can be quantified with a partial equilibrium model of the market [13]. The alternative use most likely to be affected can be identified through an econometric analysis [14].

An investment in a relatively new energy technology can contribute to improvements in that technology and thus to more such investments being made in the future. Such an indirect effect can in some cases be very large [15]. In an advanced CLCA, the effect could be roughly estimated using an energy system model with so-called experience curves [16].

An ALCA, in contrast, does not include environmental benefits or other indirect consequences that arise outside the life cycle of the investigated product. Instead, the raw material use and emissions of a co-production process are partitioned between the products of that process. In the cogeneration example above, the environmental burdens of the CHP plant are divided between the electricity and the heat. Such a partitioning is called allocation and can be done in several different ways (see Section 5.1).

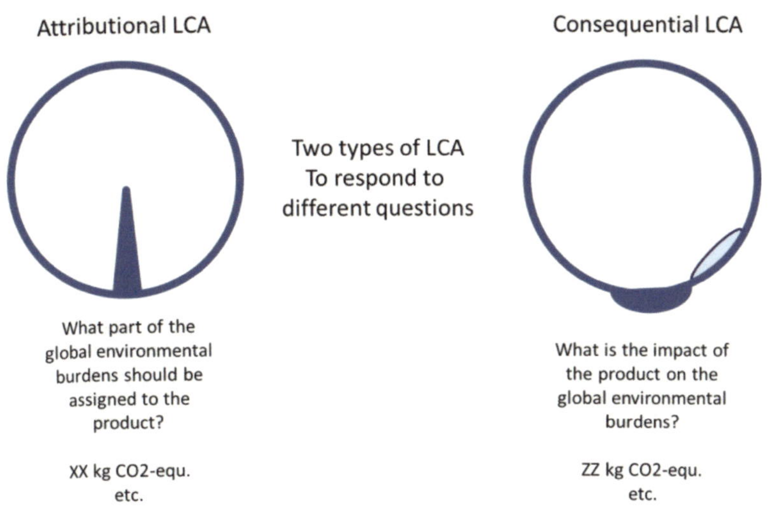

Figure 2.
Illustration of accounting and consequence LCA (based on Weidema [12]). The large circles symbolize the total environmental burdens of the world.

The choice between attributional and consequential LCA also affects the choice of input data to the calculations. An ALCA estimates how much of the world's environmental impact belongs to a product. If electricity is used in the product's life cycle, the calculations must include the product's share of the environmental burdens of the electricity production system. This is calculated by multiplying the product's electricity consumption by the average environmental burden of the electricity system per unit of electricity delivered. The figures describing the average environmental burdens are called average data. The electricity described by these average data is called average electricity.

Average data is used not only to model electricity production in ALCA. If the product investigated contains steel, average data is used to model steel production. The same applies to other input goods. In order to calculate the average environmental impact of a production system, the boundaries of the production system must be defined. This can also be done in different ways (see Section 4.1).

A CLCA aims to generate information on how the study object affects the environmental burdens of the world. If electricity is used in the system investigated, the CLCA should include data that reflects how the environmental burdens of the electricity production system are affected by this electricity use. In a few cases, the system investigated has a significant impact on the electricity production—for example, in a study of a future electric car fleet. In such cases, the CLCA should ideally be based on input data that reflects how such a large change in production volume would affect the production system's environmental burdens. Such data are called incremental data. With incremental data, the environmental burdens per kWh electricity often depend on the size of the change in power generation (compare the slope of the two lines representing incremental data in **Figure 3**).

In most cases, however, the electricity use in the system investigated is so small it has only a marginal impact on the electricity system. A change can be described as marginal when it occurs within a range where the environmental burdens as a function of the production interval can be approximated with a straight line (see **Figure 3**). Within this range, the slope of the line represents the approximate increase in environmental burdens per unit increase of electricity produced. Since the line is straight, the environmental impact per kWh is approximately constant, and the environmental impact of an additional electricity demand is proportional to the size of this demand. Data that reflects the environmental impact per kWh change within this range is called marginal data. The electricity described by marginal data is called marginal electricity.

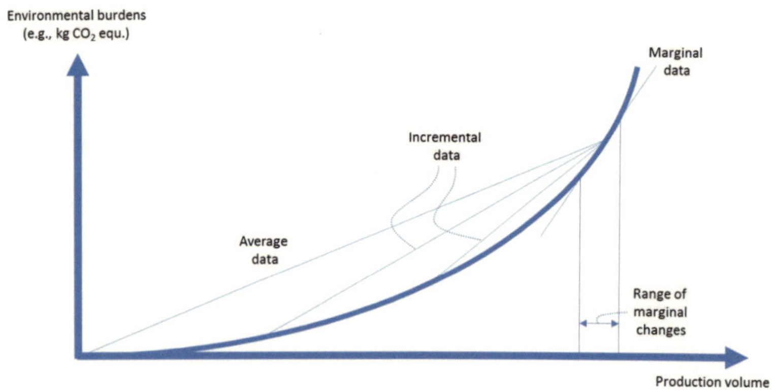

Figure 3.
Illustration of average data, incremental data, and marginal data (based on Azapagic and Clift [17]).

A CLCA should, if possible, include marginal data not only on electricity production but also on the production of other inputs where the study object only has a marginal impact on the total production volume. There are different types of marginal effects and different ways of identifying marginal production. This is further discussed in Section 4.2.

A CLCA can be made to describe and estimate the consequences of a given decision but also to investigate what a specific decision-maker can influence. If this decision-maker can completely shut down or replace a production system, the CLCA should include the entire production system. The environmental burdens per unit produced in this system are then the total burdens of the system divided by the total product output. This is identical to the average data.

4. The choice of average and marginal data

If marginal or average data are to be used in the LCA depends on whether the study is attributional or consequential, as discussed above. However, there are several types of average and marginal data. The next question to ask is therefore what average or marginal values should be used as input in the calculations.

4.1 The average of what?

An ALCA is based on average data on the production systems in the product life cycle. In order to calculate the average environmental impact of the production systems, they must be identified, and their boundaries must be defined.

When the supplier of a material or component is known, this supplier is linked to the product through contracts and through the economic and physical flows resulting from the contracts. Established good ALCA practice is then to use as specific data as possible. These are data representing the average environmental performance of the supplier or, when possible, of the individual processes in the production plant.

In many cases the supplier is unknown, for example, because the product is not yet being produced or because the material or component is bought on a market where the actual supplier shifts over time. Here, established ALCA practice is to use average data for the relevant geographical area. Ideally, this is the area from where the good is bought and/or the area covered by the market, which might be global or regional.

Energy carriers like electricity, gas, or district heat are distributed in networks. When the suppliers are known, there are contractual links and economic flows to the supplier, but there is no clear physical flow from the production process to the user. If the contract specifies the producer, it is rather uncontroversial to use data representing a weighted average over the production plants that the supplier has in the network.

Contracts might also specify that the electricity bought is produced with a specific technology, such as wind power. In such cases, it is reasonable to use data for wind power in the ALCA. To be more specific, it is reasonable to use average data for the wind power of the producer or supplier to which the contract applies. If the deal is on wind power from a specific plant or site, average values for that plant/site should ideally be used. Of course, similar rules apply if the contract specifies that the electricity is hydro or some other specific technology, or green electricity in general.

When the electricity supplier is unknown, many influential LCA guidelines (e.g., [18–20]) recommend the use of national average data or, for very large countries, average data for regional electricity grids. This might be because electricity supply has traditionally been a responsibility of national authorities. For the past decades, electricity production has been privatized in many countries, power producers have become international companies (e.g., EDF, Vattenfall, E.ON), electricity grids have become more integrated nationally and between countries, and electricity trade and transfer between countries have increased. This means that most electricity systems are no longer isolated national or regional grids. There are strong arguments for using average data for a larger geographical area instead. However, there are various ways to define this area. I here discuss them with a focus on Northern Europe, where I have my expertise:

Although production of electricity is increasingly privatized, the electricity sector is still to a large extent regulated by national authorities. One way to defend the use of national average data is to define the electricity system by the geographical scope of regulating authorities. Note, though, that electricity production is affected not only by national authorities but also by local authorities and by international cooperation, for example, within the European Union (EU).

Another approach is to define the geographical area by the electricity market. Since the establishment of the Nordic electricity exchange, NordPool, there is a well-established Nordic market, and the corresponding electricity system is often perceived as Nordic, including Sweden, Norway, Denmark, and Finland. As NordPool expands and the transmission capacity to other parts of northern Europe increases, it becomes increasingly relevant to regard the market as North European. There is also an EU directive aiming toward a common European electricity market, with provisions to remove bottlenecks in the electricity transfer between countries. In the future, the electricity market may be described as pan-European.

The electricity system can also be defined based on physical facts, for example, the transmission capacity between or within countries. This can be insufficient at times when a lot of electricity is produced at one place and used elsewhere. As a result, there will often be a difference in electricity price, for example, between North and South Sweden and between North and South Germany. The boundaries of the system can be defined where the transfer of electricity is limited by the transfer capacity in the grid, for example, between northern and southern Germany.

Alternatively, the electricity system can be defined as the area where the electricity network is synchronized, allowing for transfer of electricity without conversion to direct current. Conversion of electricity is a bottleneck because it is associated with energy loss. Based on this physical bottleneck, a system boundary is between Jutland and Zealand in Denmark, where the former is synchronized with continental Europe but the latter with the rest of Scandinavia.

Regardless of the geographical boundaries of the electricity system, the question remains as to whether data should apply to the average of the electricity produced in this area or whether they should apply to the average of the energy used in the area. In the latter case, imports and exports of electricity must be accounted for in the calculation of the average.

4.2 What marginal impacts?

The difference between short- and long-term marginal effects is important in a CLCA [13]. The distinction between short and long term is well-established within economic theory. Short-term effects in economics are effects on the utilization of existing production capacity that occurs before the production capacity has been able to adapt to, for example, a change in demand. The capacity itself is thus assumed to be unaffected in a short-term perspective.

When long-term effects are examined, the production capacity is assumed to completely adapt to the change in demand, to the extent that the risk of capacity shortage is the same as before the change. For the production of most goods, this means that the utilization rate of the capacity is assumed not to change. However, for electricity the long-term marginal effect of increased electricity use may include the construction of, for example, new wind turbines that have lower utilization rates than other power plants. This reduces the total utilization rate in the electricity system, although the risk of capacity shortage is unchanged.

If the electricity use in the life cycle is small, the probability is very small that it will affect the energy system's production capacity. Electricity for lighting in a single house is, for example, a drop in the sea, compared to the total production capacity of the electricity system. The sea, on the other hand, does not consist of much else than drops. If a change in the lighting of a house happens to be what triggers an investment in a new power plant, the effect of the lighting becomes much greater than the electricity demand of the lighting. The long-term marginal effect is calculated as the expected value, i.e., the small probability times the large outcome. This expected value is 1 kWh/year changed production capacity per kWh/year change in the consumption of electricity.

The short- and long-term marginal effects can be difficult to communicate, as they are easily confused with the effects of changes made in the near or far future. However, short-term effects can arise far into the future, and long-term effects can occur in the coming decades. As an example, the long-term marginal technologies in 2020 are the technologies whose production capacity is affected by energy use in 2020. These effects may occur in 2025–2035. Meanwhile, the short-term marginal effects in 2050 relate to how a change in energy use in 2050 affects the utilization of the production facilities that exist in 2050. These effects occur during that same year and the years immediately thereafter. Short-term marginal effects of a disruption in 2050 thus arise later than the long-term effects of a disruption in 2020.

To make communication easier, the concepts short- and long-term marginal effects are sometimes replaced by "operating" and "built" margins. A draw-back of this terminology is that the term built margin is somewhat misleading: changes in production capacity are not always the construction of new facilities; it may instead be the closure of existing production facilities. The long-term marginal effects of a change in energy use in the year 2020 can include technologies in energy plants that are constructed during the period 2025–2035, but they can also include technologies in energy plants that are shut down during the years 2020–2030.

Which concepts to use depends on the context. In communication with the general public, the rough meaning of the concepts should be easily understood. Operating and built margin are good terms to use in this context. In communication with researchers in the field, however, the precision of the concepts is important. Then it is probably better to talk about short- and long-term effects. In communication with policy-makers and professional actors in the industry, the appropriate choice of words may depend on the situation and the level of knowledge of the audience.

Changing demand for a product often gives rise to both short- and long-term marginal effects: the utilization rate is affected first, and after a while the change also contributes to new power plants being built or old ones being shut down. Changing demand can also affect investments in several different technologies, and these investments can in turn affect both the utilization rate of existing plants and other, future investments. This means that the full marginal effect is complex. The complex margin in an energy system can be estimated in an optimizing, dynamic model that can account for both the short-term and long-term margin changes [21]. The complex marginal effect is then defined as the difference between the results of two model runs: one with the change in energy demand and one without it.

The complex margin is, in theory, the most correct to use for CLCAs whether the possible decisions involve changes in the short term (e.g., putting out a lamp) or the long term (e.g., changing the heating system in the house). This is because even short-term changes can produce long-term marginal effects. Investment decisions are based on assessments of the future demand and price of the product.

These assessments are, in turn, affected by the current market situation. If we increase electricity consumption this year, we might contribute to investment decisions being taken next year or the year after that.

In practice, the complex marginal effects are very difficult to estimate. It requires model calculations over the relevant time period. Model runs suggest that this time period never ends, because indirect effects occur when new production plants must be replaced far into the future [21]. Unfortunately, the uncertainty very far into the future is too great for modeling to be meaningful. The choice of time horizon in the model is subjective and depends on the time resolution in the model. If each year is modeled as a single or a handful of time slots, the model usually extends a couple or a few decades into the future [21–24]. An hour-by-hour model is more likely to cover just a single year [25], although it can still be possible to model a few years where each model year represents, for example, a decade [26].

Identifying marginal effects with an energy system model requires special expertise. There are rarely resources to develop an energy system model within the framework of a specific LCA. With the right expertise, the marginal effects can be studied in an existing model. It is, of course, even easier to use results from published model runs as a basis for assumptions about the marginal effects. Assumptions about marginal effects of electricity use in Sweden can be based on results from, for example, Hagberg et al. [26]. However, the simpler the method used to generate complex marginal data, the greater the risk that they do not reflect the marginal effects caused by the specific electricity use being studied.

Perhaps the biggest problem is that the uncertainty in complex marginal data is extremely large. Optimizing dynamic energy systems models indicate that the complex marginal effects of Swedish electricity use vary greatly depending on assumptions on, for example, investment costs, future fuel prices and policy instruments—where the two latter are highly uncertain [21]. Completely different marginal effects can occur in a single electricity scenario, depending on whether the expansion of wind power in the scenario is assumed to be driven by an increased electricity demand or by other motives [26]. A small change in the use of district heating can change the optimum development of an entire district heating system completely [24]. This illustrates that the actual effects of a small change in demand are and will remain basically unknown. An optimizing dynamic systems model can remind us of the great uncertainty, but not give much knowledge of the actual marginal effects.

Referring to the criteria in Section 2, input data on complex marginal effects make the CLCA results more accurate, but just a little—particularly if these data are from previously published model runs. Generating case-specific complex marginal data leads to a method that is difficult to use. The use of complex marginal data also makes the study less comprehensible: it is a challenge to explain marginal results from an energy system model. This makes it more difficult for decision-makers to assess the relevance and validity of the results.

If complex marginal effects are to be introduced at all in a CLCA depends on the context. In many cases, it is probably better to use a method that is easier to use and explain. The LCA practitioner and the decision-makers should then be aware that the method used is simplified and that the actual marginal effects remain unknown.

A simplified method can be limited to focusing on short- or long-term marginal effects only. Since investment and closure decisions have consequences for the environment during a long time, such effects are typically more important for the environment than changes in the use of existing production capacity. In other words, the long-term marginal effects are typically more important for the environment than the short-term marginal effects [13].

In some cases, however, a change in demand cannot be expected to have any effect at all on the production capacity. This applies if the existing production system has a significant overcapacity and closure of existing plants is not a reasonable option. It also applies if the production capacity is expanded for political or other strategic reasons, rather than to cover an expected demand for the product. A change in current Swedish electricity use might, for example, not have any effect on new investment decisions, because there is an overcapacity in the North European electricity system and because wind and solar power is still being expanded for policy and strategic business reasons. On the other hand, a change in electricity use can contribute to keeping electricity prices up or down, which can make decisions on continued investments more or less difficult. There is also a long-term political ambition to phase out coal and nuclear power. A change in electricity demand can contribute to a quicker or slower closure of such power plants. This discussion reminds us that the actual marginal effects are difficult to foresee. Different assumptions are possible, even if the environmental assessment is limited to long-term marginal effects.

Another way to simplify things is to use the five-step procedure presented by Weidema et al. [27] to identify the production technology that is affected by a marginal change in demand. This procedure involves responding to five questions:

1. Is short or long term the relevant time perspective?

2. What market is affected? Here, both a geographical delimitation and a delimitation in different market segments may be required, for example, in base- and peak-load electricity or in eco-labeled and non-ecolabel products.

3. What is the trend in demand in this market? If demand declines faster than the natural turnover rate in production capacity, long-term marginal effects are assumed to consist of closure of existing plants; otherwise they are assumed to consist of investments in new facilities.

4. Which production techniques are flexible, that is, can vary their production volume in response to market demand?

5. Which technology will be affected? If the marginal effect is an investment, it is assumed to be in the technology that is cheapest to expand. If the marginal effect is a closure, it is assumed that it is in the technology that is most expensive to utilize.

This five-step procedure can be used in CLCAs of a wide range of products. The procedure points at a single technology where the marginal effect occurs. This contributes to making the CLCA approach feasible and comprehensible—but at the cost of simplifying assumptions: that the relevant effects are either short-term or long-term rather than both, that markets and market segments can be clearly distinguished and do not affect each other, that the production volume of a technology is either completely flexible or not at all flexible, and that decisions are based solely on economic rationality. Each of these simplifications reduces the accuracy of the CLCA results. The LCA practitioner and the user of the LCA results should both be aware of this. The five-step procedure can be described as a structured way to arrive at an assumption of the marginal effects, rather than a method of identifying the actual marginal effects.

Another approach is to collect information on plans to close and/or expand the production capacity and assume that the built margin is the mix of technologies in these plans. This is also an assumption, because plans do not always come true [28] and because some of the closure and investment decisions might be driven by policy or business strategies rather than by the demand for the product.

Assumptions about the marginal effects can, of course, be made even without a structured or formal procedure. Long-term marginal effects in the electricity system can, as the first approximation, be assumed to be electricity production in new natural gas-fired power plants, as they have an environmental performance that is better than some possible marginal techniques but worse than others. A possible sensitivity analysis can be based on data from old coal power or old nuclear power, as the closure of such power plants can be included in the long-term marginal effects and because they are near opposite ends of the scale for several important environmental impacts. Similarly, a first approximation and the extreme values can be identified for marginal production of other products.

To simply make an assumption is likely to be the easiest method to produce marginal data for the environmental assessment. On the other hand, pure assumptions make the study less accurate. They can also make the study less comprehensible in the sense that the basis for the assumptions can be difficult to communicate. If the assumptions appear arbitrary, the study also becomes less credible, which reduces the likelihood that the results inspire decisions and actions.

5. Dealing with allocation

A single production process often serves many different life cycles: diesel from a single refinery and steel from a steel mill can be used in almost any life cycle. If the production process generates a single type of product (e.g., steel), this is not considered a problem in LCA. We obtain input data to the calculations by simply dividing the total environmental burdens of the process by the total production, the functional output, of the process. The resulting input data are an average for that process and, hence, most suited for an ALCA. In a thorough CLCA we should ideally instead use input data that reflect how the environmental burdens of the process change as a result of a change in the total functional output. This is still a straightforward process, at least in theory.

A methodological problem occurs when the process generates more than one type of product or function, which are used in different life cycles. A refinery, for example, produces many different fuels and materials. A steel mill might produce residual heat besides the steel. A CHP plant produces electricity and heat. Waste incineration serves the function of treating many different waste flows and might, at the same time, generate electricity, residential heating, and/or process steam. The problem is to decide on how quantify the total functional output of the multifunctional process and, hence, how to allocate the environmental burdens of the process to the various life cycles it serves. The approach to this problem depends on whether the LCA is an ALCA or a CLCA.

5.1 Partitioning in attributional LCA

An ALCA aims to estimate what share of the global environmental burdens belongs to the product investigated. Faced with the allocation problem, the task is to estimate what share of the burdens of the multifunctional process belongs to the product investigated and also what share of input materials, energy, etc. The basis for this allocation has to be a property that the products and/or functions of the process have in common: mass, energy content, economic value, etc. The total output of the process can be quantified in terms of this property, and the burdens of the process can be partitioned and allocated to the different products/functions in proportion to this property.

What properties the products and functions have in common varies between multifunctional processes:

- A refinery: mass, energy, exergy, and price
- A CHP plant producing electricity and heat: energy, exergy, and price
- A steel mill with residual heat: price
- Waste incineration with energy recovery: price

As indicated from this short list, the price is sometimes the only possible basis for allocation. In many ALCAs, it is the only allocation key that can be consistently used throughout the life cycle. Economic value can also be considered a valid basis for the allocation, since the economic value of the products is a proxy for their contribution to the expected profit from the process. The expected profit is typically the reason for investing and running the process and, hence, the cause of its impacts on the environment [12, 29].

Economic allocation is often criticized because it will make the LCA results vary as prices change over time. However, the LCA results can be made more stable by using the average price over a period of several years as basis for the allocation. This will also more precisely reflect the causality, because the expected profit is more likely to depend on the average price than on the price at a specific point in time.

There are cases where the economic value does not reflect a causality, because the processes are not driven by the expected profit but by concern for, e.g., the environment. These include noncommercial processes such as municipal wastewater treatment plants [30] and landfills. In these cases, the economic value is less valid and might not even be possible to use as basis for the allocation.

When we choose an allocation key, we might account for what the intended audience considers to be fair. This increases the legitimacy of the study in their eyes, which increases the chance of the LCA leading to decisions.

The choice of allocation method also depends on how feasible it is. If the allocation problem is not important for the results and conclusions of the ALCA, the easiest methods can be used to keep the cost of the study down. This can include allocating all burdens to the main product of the process—for example, to the steel from the steel mill with residual heat.

5.2 System expansion in consequential LCA

A CLCA aims to estimate how the global environmental burdens are affected by the production and use of the product investigated. Faced with a multifunctional process, the task is to estimate how the flows of the process are affected: the flows of input materials and energy, the emissions and waste flows, and the output of each product and function. When the output of products and functions for use in other life cycles are affected, the CLCA system should ideally be expanded to include the processes that are affected by this change in flows.

A change in the demand for one of the products from a multifunctional process can affect decision-makers running the process and other actors in various ways that are difficult to predict and model. To make the CLCA approach feasible, we can choose to divide the multifunctional processes into three idealized cases [13]:

1. Independent production: a change in the demand for the product investigated affects the output of this product but not the flow of other products and functions from the process.

2. Use of main product in joint production: an increase in the demand for the product investigated drives the process and increases the output of all its products and functions proportionally.

3. Use of by-product in joint production: a change in the demand for the product investigated does not affect the process or any of its outputs; instead it affects the alternative use of the by-product.

The idealized cases are simplifications of reality: products from a multifunctional process are rarely produced completely independent of each other [31], and the process is rarely driven by only one of the functional outputs.

If the products of the multifunctional process are independently produced, the input data for each of the products should reflect how the environmental burdens of the process change when the production of this product changes while the production volume is constant for the other products.

If the CLCA includes the use of the main product from a joint multifunctional process, the LCA model should include this process and also the processes affected by a change in the volume of by-products. The latter are typically assumed to be the production of products that compete with and are substituted by by-products from the multifunctional process (see **Figure 4**). Since the study is a CLCA, the competing production should ideally be modeled based on marginal data (cf. Sections 3 and 4.2).

If the CLCA instead includes the use of a by-product, the operation of the multifunctional process is assumed to be unaffected by the demand for this product. The use of such a by-product does not affect its production; instead, it affects how much of the by-product is available for other purposes. The CLCA model should include affected processes only, which means it should not include the multifunctional process. Instead, the model ideally includes the marginal, alternative use of the

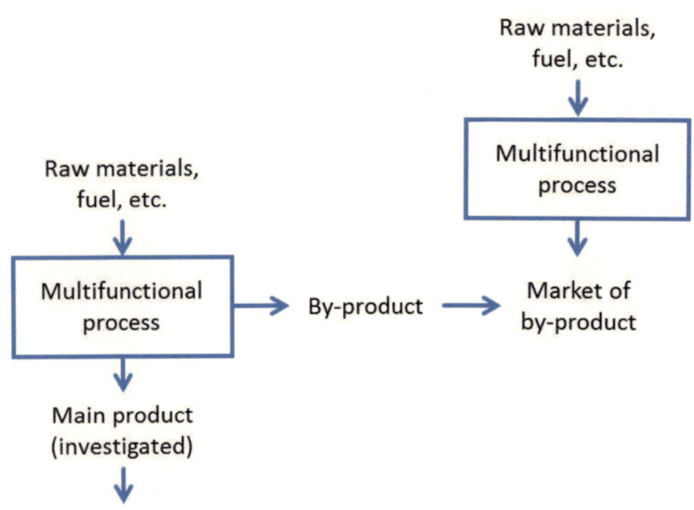

Figure 4.
System expansion at a joint multifunctional process where the product investigated is the main product (based on Ekvall and Weidema [13]).

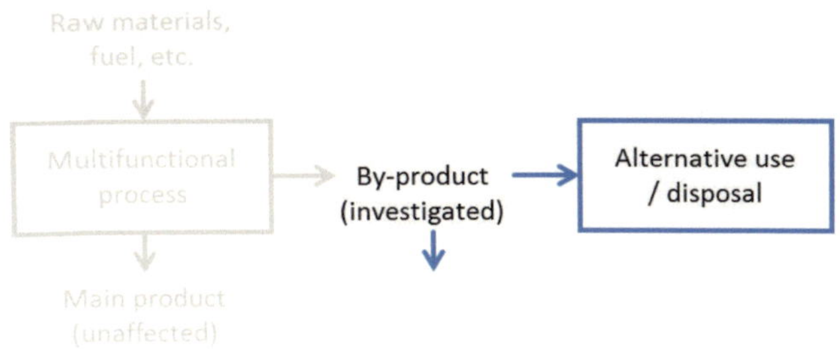

Figure 5.
System expansion at a joint multifunctional process where the product investigated is a by-product (based on Ekvall and Weidema [13]).

by-product. This is the use affected by a (usually marginal) change in supply of the by-product (**Figure 5**).

In some cases, the by-products are not fully utilized: for example, part of the residual heat from a steel mill might be cooled off, and part of a residual material might be disposed as waste. In such cases, a change in the use of the by-product is not likely to affect the alternative use but instead how much of the by-product needs to be cooled off or disposed of in some other way. The CLCA should include the affected disposal process.

Note that the "expanded" system in **Figure 5** is not necessarily larger than the original system. It does not include the multifunctional process or the production of fuel and other raw materials for that process. Instead, it includes the disposal or alternative use of the by-product and any foreseeable consequences thereof.

The easiest method, such as ignoring the production of by-products, can be applied in the CLCA if the choice of approach is not important for the results and conclusions of the study. However, more information is required to decide on such a cut-off in a CLCA, compared to an ALCA. Even if a multifunctional process has little environmental burdens, making it unimportant in an ALCA, a change in this process might have environmentally important consequences elsewhere, hence making it significant for the CLCA.

6. The pros and cons of attributional and consequential LCA

Attributional and consequential LCA have both advantages and disadvantages [9, 32]. This section discusses the choice between ALCA and CLCA using the criteria described in Section 2. The intention is not to determine what kind of LCA is superior but to discuss and explain their strong and weak aspects. The intention is also to show how the criteria in Section 2 can be used systematically to structure a discussion and assessment of methodological options.

6.1 Feasible

In a CLCA, the system model often needs to be expanded (Section 5.2), which requires environmental data on more processes and also economic data on the markets affected by the production and use of the product investigated (cf. Section 4.2). The databases that exist today usually include average data, but few include marginal data—Ecoinvent 3 is a notable exception, although its marginal data are rough. All of this means that a CLCA risks becoming unfeasible or at least significantly more expensive than an ALCA. On the other hand, the CLCA can exclude parts of the life cycle that are not affected by the production of by-products (cf. Section 5.2). The cost of CLCAs can also be reduced by limiting the study to the consequences expected to be the most important for the conclusions.

With time, CLCAs may become easier to carry through if future databases include more of marginal data.

6.2 Accurate

A CLCA generates information on the environmental impact of a specific decision or information on how a decision-maker can affect the environment. This is just the accurate information to have as a basis for decisions that contribute to reducing the total negative environmental impacts or, at least, the impact per functional unit.

An ALCA might be more precise and comprehensive, because a detailed and comprehensive CLCA might be too expensive or even unfeasible to carry through (see Section 6.1). As an ALCA is refined, it becomes more detailed, and the results converge toward an exact response to the attributional question: how much of the world's environmental impact belongs to the product studied? However, even a very precise answer to this question will in some cases guide decisions in the wrong direction, because the impacts belonging to a product are not the same as the consequences of producing and using this product (see **Figure 2**).

Refining a CLCA can involve accounting for more causal relationships. This makes the CLCA more comprehensive, but it does not necessarily mean that the results converge toward a final answer. On the contrary, as an additional causal relationship is included in the calculations, the results might shift completely and point in another direction.

The CLCA provides, by definition, more information on how decisions affect the environment; however, if the CLCA results are highly uncertain and do not converge toward a final, true result, the CLCA might not guide decisions in the right direction more often than an ALCA.

6.3 Comprehensible

An ALCA is based on the concepts "life cycle" and "value chain" which are intuitively clear and easy to communicate. The system model in an ALCA usually has a simple structure, which means that it can easily be presented in a way that is transparent, at least in principle. The high level of detail that can be achieved, however, makes the study bulky and can make it a challenge in practice to communicate to decision-makers and other stakeholders.

The basic concept in a CLCA is "consequences." This is also intuitively easy to understand. However, other concepts required to understand the study (marginal production, partial equilibrium, etc.) are more difficult to grasp. The system model is also more complex with environmental burdens, avoided burdens, and additional, indirect burdens and with models of markets between the models of production processes. Making such a study comprehensible to decision-makers and stakeholders can be very difficult.

6.4 Inspiring

An ALCA can be interpreted to distribute responsibility and guilt for environmental impact, and recognition and goodwill for environmental improvements in the value chain, a part of the technological system that is linked to the production and use of the product through contracts and/or physical flows. An LCA model based on such clear links can be perceived as a relevant basis for choosing between products and for decisions on changes in the product. If the choice of allocation methods and system boundaries is accepted by the decision-maker, the results will also be perceived as fair and legitimate. However, they can be questioned by actors who have other, subjective perspectives on what is fair and right.

The fact that a CLCA provides information on how possible decisions affect the environment can also be perceived as very relevant to the decision-maker. Rational decision-making requires information on the consequences of the decision. However, the CLCA typically include indirect consequences occurring in processes to which the product is not linked through physical flows or contractual obligations. The decision-maker might not want to be held responsible for such consequences. In order to account for them anyway, the decision-maker probably needs to be driven by the desire to actually improve the environment, rather than simply getting recognition for good environmental performance.

6.5 Robust

The ALCA practice is more well-established than CLCA. Environmental product declarations, a specific application of ALCA, also have detailed guidelines specifying the method [19]. In other applications, ALCA requires subjective choices of system boundaries (Section 4.1) and allocation methods (Section 5.1). However, the ALCA results are somewhat less sensitive to subjective choices than CLCA where the results might shift from positive to negative depending on system boundaries and assumptions. All this implies that ALCA is more robust and more resistant to abuse in the sense that the results depend less on who is doing the study.

The actual consequences of a decision are almost always highly uncertain. If the sensitivity analysis of a CLCA takes full account of the great uncertainty, the study will rarely reach clear conclusions. This increases the risk of decisions and actions not being taken, especially if the actions are expensive or undesirable in other ways. The large uncertainty in the actual consequences makes it easy to misuse CLCA results to cast in doubt environmentally desirable decisions.

However, when the ALCA is completed, the results can be abused if presented as a basis for decisions. This is because the ALCA does not aim to investigate the consequences of the decision on the environment. In a country with little fossil-based power production, such as Norway or Sweden, an ALCA can, for example, conclude that energy efficiency is not important for electric appliances. It can also indicate that residential heating should be provided through heat pumps rather than district heating from CHP plants fired with natural gas and perhaps even biofuel. A CLCA would not be likely to produce such results. If and when CLCA practice becomes more established, it will also become somewhat more difficult to abuse.

7. Conclusions

Attributional and consequential LCA respond to different questions: what part of the global environmental impacts is associated with the product investigated, and how does the product affect the global environmental impacts? In most applications and for most study objects, the choice between ALCA and CLCA is open. Since the two types of LCA have different advantages and disadvantages, it cannot be unequivocally stated that one is better than the other [32]. Roughly stated, the CLCA is more accurate, while ALCA have advantages when it comes to all other criteria. However, what kind of study is easiest to understand and most inspiring will vary between different decision-makers.

To ensure that the study is perceived as relevant, it is a good idea to, if possible, discuss the goal and scope of the study with the client before deciding on what type of LCA to carry through. To make the study as legitimate as possible, it might also be useful to discuss with other stakeholders. In such a discussion, it is important to carefully explain what type of information is provided by an ALCA and a CLCA. **Figure 2** can be used in that explanation. It is also important to make clear the limitations of the different methods. Only then can the client and other stakeholders decide on the type of study they want.

As should be clear from this chapter, the actual effects of a decision on the global environmental impacts are in most cases highly uncertain. We will never know how close the CLCA results are to reflect the actual consequences. For this reason, CLCA should probably not be presented as a method to estimate the actual consequences. Instead the results are the consequences foreseeable within the methodological framework we choose to use in the study.

The risk that the study will be abused will also vary from case to case. Here, it does not help to consult the client. The LCA practitioners must instead use their own judgment and decide what kind of LCA is the most appropriate. In this decision, it may be good to consult with colleagues and/or to discuss with other stakeholders. The decision to make an ALCA or a CLCA should therefore be taken by the LCA practitioner after discussions with the client and possibly with other stakeholders and colleagues.

Acknowledgements

This chapter was completed with financial support through the EU ERA-Net Sumforest project BenchValue (Formas Grant Number 2016-02113). It is based on translated parts of a report in Swedish, "Miljöbedömning av energibärare – vägledning för livscykelanalyser," produced within the framework of the North European Energy Perspectives Project (NEPP). Jenny Gode, Mathias Gustavsson, Erik Dotzauer, and Lisa Hallberg contributed valuable input to this report. The methods and concepts presented in this chapter were to a large extent developed by or in collaboration with Bo Weidema, Göran Finnveden, and other international LCA experts. However, the author is solely responsible for the final content of the chapter.

Author details

Tomas Ekvall
IVL Swedish Environmental Research Institute, and Chalmers University of Technology, Gothenburg, Sweden

*Address all correspondence to: tomas.ekvall@ivl.se

References

[1] ISO 14040:2006. Environmental Management – Life Cycle Assessment – Principles and Framework. Geneva: International Organisation for Standardization; 2006

[2] ISO 14067:2018. Greenhouse Gases — Carbon Footprint of Products — Requirements and Guidelines for Quantification. Geneva: International Organisation for Standardization; 2018

[3] Ekvall T, Ciroth A, Hofstetter P, Norris G. Comparative assessment of attributional and consequential methods for LCA. Poster and handout. In: Abstracts 14th Annual Meeting of SETAC-Europe; April 2004. Prague. Brussels: SETAC-Europe; 2004. p. 197

[4] Heintz B, Baisnée P-F. System boundaries. In: Life Cycle Assessment. Workshop Report: 2-3 December 1991; Leiden. Brussels: SETAC-Europe; 1992. pp. 35-52

[5] Weidema BP. Development of a method for product life cycle assessment with special references to food products [PhD thesis summary]. Lyngby: Technical University of Denmark; 1993

[6] Tillman A-M. Significance of decision-making for LCA methodology. Environmental Impact Assessment Review. 2000;**20**:113-123. DOI: 10.1016/S0195-9255(99)00035-9

[7] Ekvall T. System expansion and allocation in life cycle assessment – With applications for wastepaper management [PhD thesis]. Gothenburg: Chalmers University of Technology; 1999

[8] Curran MA, Mann M, Norris G. The international workshop on electricity data for life cycle inventories. Journal of Cleaner Production. 2005;**13**(8):853-862. DOI: 10.1016/j.jclepro.2002.03.001

[9] Zamagni A, Guinée J, Heijungs R, Masoni P, Raggi A. Lights and shadows in consequential LCA. International Journal of Life Cycle Assessment. 2012;**17**:904-918. DOI: 10.1007/s11367-012-0423-x

[10] Ekvall T, Azapagic A, Finnveden G, Rydberg T, Weidema BP, Zamagni A. Attributional and consequential LCA in the ILCD handbook. International Journal of Life Cycle Assessment. 2016;**21**:293-296. DOI: 10.1007/s11367-015-1026-0

[11] Finnveden G, Hauschild MZ, Ekvall T, Guinée J, Heijungs R, Hellweg S, et al. Recent developments in life cycle assessment. Journal of Environmental Management. 2009;**91**(1):1-21. DOI: 10.1016/j.jenvman.2009.06.018

[12] Weidema BP. Market Information in Life Cycle Assessment. Environmental Project no. 863. Copenhagen: Danish Environmental Protection Agency; 2003. 147 p. Available from: https://www2.mst.dk/Udgiv/publications/2003/87-7972-991-6/pdf/87-7972-992-4.pdf [Accessed: 2019-08-11]

[13] Ekvall T, Weidema BP. System boundaries and input data in consequential life cycle inventory analysis. International Journal of Life Cycle Assessment. 2004;**9**(3):161-171. DOI: 10.1007/BF02994190

[14] Ekvall T, Andrae A. Attributional and consequential environmental assessment of the shift to lead-free solders. International Journal of Life Cycle Assessment. 2006;**11**(5):344-353. DOI: 10.1065/lca2005.05.208

[15] Sandén BA, Karlström M. Positive and negative feedback in consequential life-cycle assessment. Journal of Cleaner Production. 2007;**15**(15):1469-1481. DOI: 10.1016/j.jclepro.2006.03.005

[16] Mattsson N. Internalizing technological development in energy systems models [LicEng thesis]. Gothenburg: Chalmers University of Technology; 1997

[17] Azapagic A, Clift R. Allocation of environmental burdens in multiple-function systems. Journal of Cleaner Production. 1999;7(2):101-119. DOI: 10.1016/S0959-6526(98)00046-8

[18] Greenhouse Gas Protocol. Product Life Cycle Accounting and Reporting Standard. Washington, DC, Geneva: World Resources Institute, World Business Council for Sustainable Development; 2011. 148 p. ISBN 978-1-56973-773-6. Available from: https://ghgprotocol.org/sites/default/files/standards/Product-Life-Cycle-Accounting-Reporting-Standard_041613.pdf [Accessed: 2019-08-11]

[19] General Programme Instructions for the International EPD® System. Version 3.0. Stockholm: EPD International AB; 2017. 77 p. Available from: https://www.environdec.com/contentassets/95ee92 11a9614f1faa7461ff32cecc91/general-programme-instructions-v3.0.pdf [Accessed: 2019-08-11]

[20] Product Environmental Footprint Category Rules Guidance: Version 6.3 – May 2018. Brussels: European Commission; 2018. 238 p. Available from: http://ec.europa.eu/environment/eussd/smgp/pdf/PEFCR_guidance_v6.3.pdf [Accessed: 2019-08-11]

[21] Mattsson N, Unger T, Ekvall T. Effects of perturbations in a dynamic system – The case of Nordic power production. In: Unger T, editor. Common energy and climate strategies for the Nordic countries – A model analysis [PhD thesis]. Gothenburg: Chalmers University of Technology; 2003

[22] Sköldberg H, Unger T. Effekter av förändrad elanvändning/elproduktion – Modellberäkningar. Elforsk Rapport 08:30. Elforsk: Stockholm; 2008. 68 p

[23] Sandvall AF, Börjesson M, Ekvall T, Ahlgren EO. Modelling environmental and energy system impacts of large-scale excess heat utilization – A regional study. Energy. 2015;**79**:68-79. DOI: 10.1016/j.energy.2014.10.049

[24] Sandvall AF, Ahlgren EO, Ekvall T. Low-energy buildings heat supply – Modelling of energy systems and carbon emissions impacts. Energy Policy. 2017;**111**:371-382. DOI: 10.1016/j.enpol.2017.09.007

[25] Lund H, Mathiesen BV, Christensen P, Schmidt JH. Energy system analysis of marginal electricity supply in consequential LCA. International Journal of Life Cycle Assessment. 2010;**15**(3):260-271. DOI: 10.1007/s11367-010-0164-7

[26] Hagberg M, Gode J, Lätt A, Ekvall T, Adolfsson I, Martinsson F. Miljövärdering av energilösningar i byggnader etapp 2: Metod för konsekvensanalys. Rapport 2017:409. Stockholm: Energiforsk; 2017. 75 p

[27] Weidema BP, Frees N, Nielsen P. Marginal production technologies for life cycle inventories. International Journal of Life Cycle Assessment. 1999;**4**(1):48-56. DOI: 10.1007/BF02979395

[28] Mathiesen BV, Münster M, Fruergaard T. Uncertainties related to the identification of the marginal energy technology in consequential life cycle assessments. Journal of Cleaner Production. 2009;**17**:1331-1338. DOI: 10.1016/j.jclepro.2009.04.009

[29] Huppes G. A general method for allocation in LCA. In: Huppes G, Schneider F, editors. Proceedings of the European Workshop on Allocation in LCA.; February 1994; Leiden. Brussels: SETAC-Europe; 1994. pp. 74-90

[30] Heimersson S, Svanström M, Ekvall T. Opportunities of consequential and attributional modelling in life cycle assessment of wastewater and sludge management. Journal of Cleaner Production. 2019;**222**:242-251. DOI: 10.1016/j.jclepro.2019.02.248

[31] Ekvall T, Finnveden G. Allocation in ISO 14041 – A critical review. Journal of Cleaner Production. 2001;**9**(3):197-208. DOI: 10.1016/S0959-6526(00)00052-4

[32] Ekvall T, Tillman A-M, Molander S. Normative ethics and methodology for life cycle assessment. Journal of Cleaner Production. 2005;**13**(13-14):1225-1234. DOI: 10.1016/j.jclepro.2005.05.010

Chapter 5

Green and Sustainability in Software Development Lifecycle Process

Mohankumar Muthu, K. Banuroopa and S. Arunadevi

Abstract

This chapter gives an insight of GREENSOFT Model for sustainable software engineering. In today's world, computing devices are extensively by all for many purposes. They consume lots of energy even though they reduce energy consumption. Computers are used extensively while developing software. Existing software engineering models do not pay much attention to green computing that focuses on the effective use of natural resources. Sustainability of resources is the key. The GREENSOFT model of software engineering proposes a methodology in which Green IT practices are used, which will reduce the energy consumption of computers while developing software.

Keywords: software engineering, environmental informatics sustainability, ICT, Green IT, GREENSOFT model

1. Introduction

Today, software is extensively used by all of us for commercial and noncommercial purposes and it is controlled by various entities. The total energy consumption by ICT is between 5 and 9% and is going to increase annually by 6–9%. (**Figure 1**) [23]. In India, the exiting ICT facilities have the energy performance index ranging between 230 and 310 kWh/m^2/year. The greenhouse gas emissions by ICT is estimated to be 2.8% in total global emissions, but as usage of ICT can also reduce the emission rate by 15%. This calculation is on consumers who use ICT devices. But energy consumption levels can be further reduced from the current levels if we apply green software development lifecycle. The existing Software Development Life Cycle process framework is not adequate when concerned with energy consumption. This is mainly due to the lack of awareness in the Green Based Software Development life cycle process. The existing Green based software development model namely Green Reference Model is focuses on the three phases of the process called development, usage, and end of life. But the GREENSOFT model does not provide Green-based approach in entire Software Development Life Cycle phases which is a major setback of GREENSOFT model.

The existing model for Green software engineering does not consider the Requirement, Design, Coding, Testing and Implementation phases of SDLC. In order to achieve green software engineering the entire software development process is to become Green.

IntechOpen

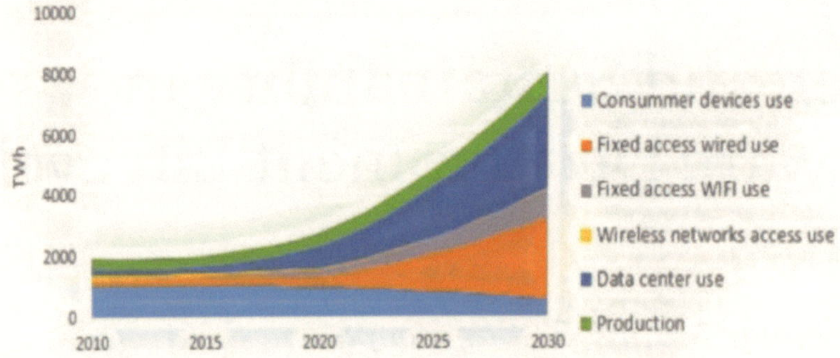

Figure 1.
ICT electricity consumption.

2. Environmental informatics

Sustainability has become a very essential factor to consider doing business in recent times. If that business has fails to maintain sustainable development it receives significant public criticism and they may lose the market [1].

The usage of computing power is to develop the more sustainable world [2]. The answer for the ecological crisis can be solved by knowledge acquisition. The development of feedback information should be concerned with model makers and system analysts. Environmental Informatics (EI) [2] is the combination of information systems of problem-oriented knowledge from the fields of management and environmental science and computer science. EI is emerged for the requirement of domain-specificity. The computer scientists have the challenge of designing the information processing system for using the environmental data.

The Computational Sustainability (CompSust) [2] is closely associated with the Institute for Computational Sustainability (ICS) with funding starting from 2008 by the U.S. National Science Foundation. CompSust is developed by ICS as "inter disciplinary field to balance the needs (environment, economical and societal) of sustainable development using computer and information science, operations research, applied mathematics, and statistics. The Brundtland definition for CompSust community refers to address the need of basic human." "Balancing" might address the issue to some extend with no reference of deeply normative issues which are connected to distributive justice. Thus, an algorithm can resolve normative issues, which is yet to be developed for greening ICT.

Greening ICT in industry application can run in design phase, classification of ICT systems development phase, execution of software systems phase and complex installation phase and in software development of the Life cycle of ICT systems [3] there are a lot of changes to save energy. The energy can be saved efficiently by using computing resources and avoiding recurring work such as recompiling. Based on World Commission on Environment and Development (WCED) [2], "sustainable development" could be defined as sustainable system use of a system to fulfill its function (F) for a time (L).

3. Sustainability

The goal of sustainability is to provide a common elucidation of the word "sustainability" without actually relating it to any specific context [2]. To perform this, they will first conclude the sustainability definitions which described as a broadly used term for the ability of something for long lasting time. Some other definitions for Sustainability are as follows.

- Define sustainability as follows "the ability to be overseen at a steady level without depleting environmental assets."
- The Brundtland report from the United Nations (UN) defines sustainable development as the ability to "meet the present needs without compromising the future generation abilities for their own needs."
- Environmental sustainability ensures that the environment is the ability to refill itself at a quicker rate than it is damaged by human actions. For instance, the use of recycled material for IT hardware production helps to conserve natural resources.

3.1 Sustainable software engineering (SSE)

Sustainable software engineering [4] motive is to create reliable, lifelong software that meets the needs of user's requirement and also tried to reducing ecological impacts; its aim is to generate better software so there is no need to compromise future generations' opportunities.

3.2 Green and sustainable software engineering

Green and sustainable software engineering [5] is the skill of creating green and sustainable software in relation to the process of green and sustainable software engineering. The skill of describing and making software products in a way, the positive and negative effects on sustainable developments of the Software Development Life Cycle process constantly assessed. The assessed details are maintained as records and may be utilized for a software product process optimization [25].

3.3 Software engineering for sustainability

The objective of software engineering for sustainability [6] (SE4S) is tried to create a tools and technique in order to reach the conception of software sustainability.

Based on [24], purpose of green and sustainable software engineering [24] is the improvement of software engineering. During the entire life cycle of software system which targets the direct and indirect consumption of natural resources and energy and the aim is to track, access, everlasting measure and optimize these realities.

SSE [7] aims to develop consistent, lifelong software that satisfies the needs of customers and also tried to reduce the negative impact on the financial, humanity and the ecological system [9]. The software engineering sustainability process tries to balance the business and technical advancement in the environment. IT (Information technology) has played a very important role to tackle issues of ecology and various types' ecological issues. However, these can be measured separately. The first consideration is to IT which could be used to mitigate ecological issues [26]. Green IT is defined as "the study and practice of design, developing, using, and disposing of computers and peripherals, and servers effectively without affecting the environment" [8]. Greening through IT [10], conversely is the center of attention on how IT may create a wider range of additional - civilization sectors for further sustainable IT application field. In a wide sense, computers are everywhere and consist of both specialized and generic systems. This report focuses on computers that constitute significant loads in buildings and specifically investigates energy-efficiency opportunities in five broad computer form factors: desktops, notebooks, small-scale servers, thin clients, and workstations. While the number of tablets in homes is increasing, the energy use of these products is relatively low, and the saving opportunity–is minimal due to existing battery charger regulations and market pressure for achieving the high efficiency in enhancing the battery life [11] which can have main impacts on economic and the social at the worldwide, which makes ecological and power issues of software worldwide concerns too. According to the Annual Energy Outlook 2017 published by the US Energy Information Administration (EIA), energy consumption is expected to increase by no more than 5% between 2016 and 2040, with the sector of electric power being the largest primary energy consumer. On the contrary, projections of energy production vary widely due to the production growth dependency on technology, resource, and market. Energy related CO_2 emissions decline in most Annual Energy Outlook scenario [12]. These categories give a sampling of the different types of technology that are being built with the purpose of enabling greening through IT. The IT segment itself is said to be accountable for 2% of global Carbon dioxide emissions [12], and the global impacts of ecology in this aspect includes high amount of energy utilization [14] and utilization of a different variety of other materials [15], making of wastes like e- and hazardous waste. The ES (Expandability Score) score of greater than 690 is considered a high expandability computer and would be subject to the standards for workstations rather than the desktop standards.

The main ecological concerns of trade are the crisis of global energy. According to World Energy Outlook 2010, "the age of cheap oil is over" [16], describing increasing power prices and for the past 30 years, while per capita electricity consumption in the United States has increased by nearly 50%, California's electricity use per capita has been nearly flat, Continued progress in cost-effective building and appliance standards and ongoing. Due to the fast increasing IT demand,, energy utilization of IT is also a needed for investigation [17] The Long-Term Energy Efficiency Strategic Plan calls on the Commission to develop a phased and accelerated "top-down" approach to more stringent codes and standards. It also calls for expanding the scope of appliance standards to plug loads; process. Computers contribute significantly to energy consumption in the commercial sector, particularly in office buildings and schools. In fact, the U.S. Energy Information Administration's analysis of miscellaneous loads suggests that 70% of commercial notebook and desktop energy consumption occurred in these types of buildings in 2011 [4, 18]. These assumptions include the rising IT role in power administration, technological developments.

3.4 Power factor

Power factor correction is important to power supply efficiency. The California (investor-owned utilities) IOUs proposed to include testing and minimum standards for power factor at full load to achieve energy savings on both the consumer side of the meter as well as on the utility side. NRDC (Natural Resources Defense Council) further recommended power factor correction at lower load points, including sleep and off, to increase energy savings [26]. It may be propose a minimum power factor requirement at full load for computers with non-federally regulated power supplies to ensure consistency with other power supply standards, including the federal external power supply standards and the 80 PLUS® program. However, requiring minimum power factor at low loads demands additional technical support to demonstrate technical feasibility and cost-effectiveness that was not available at the time of this effort [29].

3.4.1 Regulations for computers

For the potential energy savings staff have included desktop computers (including integrated desktops and portable all-in-ones), notebooks (including mobile gaming systems, two-in-one notebooks, and mobile workstations), small-scale servers, thin clients (including mobile thin clients), and workstations (including rack-mounted workstations) in the proposed regulations. A thin client is a type of desktop computer that relies on a server or networked virtual machine to provide full functionality, such as data storage and computational power. Staff have excluded other servers, tablets, smartphones, setup boxes, game consoles, handheld video game devices, small computer devices, smart televisions, and industrial computers [26, 29].

3.5 ICT for sustainability

ICT for sustainability (ICT4S) [2] means metamorphic capability of ICT can be used to make our patterns of manufacturing and exhaustion more feasible [26]. At the same time, the history of scientific knowledge has exposed that increased energy performance does not vitally contribute to green development. To create a more sustainable culture in true potential of ICT can be possible with the efforts of politics, industry and consumers.

ICT4S can be part into:

- Sustainment in ICT: accomplishing ICT products, enterprises also economical above their entire existence cycle, basically diminishing the vitality, material streams which summon

- Empowering by ICT: The method for ICT primary angle are secured by Green ICT and making authorize, empower, creation and utilization by TRHCI and EI. In the event that there is something particular to ICT4S as a field, it is the basic viewpoint that difficulties each mechanical arrangement by surveying its effect at the societal level.

3.6 Software sustainability

There are various territories in which programming manageability longings to be connected [21]: framework programming, programming related items, Web based applications, server farms, and so forth. Diverse works are in process, yet the greater part of this worries server farms, which expend extensively best vitality than business office space. As noted in, the core opportunity for energy savings in computers is found in reducing the amount of energy consumed in idle modes; that is, when the computer is on but not being used [27]. Idle modes are the largest opportunity to reduce energy consumption because computers spend roughly half of the time in this "on mode." In addition, high idle-mode consumption greatly increases the effectiveness of power management settings to reduce overall computer energy consumption. Automatic power management settings are often disabled, which means computers are constantly consuming significant amounts of power when not in use (for example, 50 W in idle mode compared to 2 W in sleep mode).

The software sustainability [30] is the one part of the software engineering sustainability should usually be considered into account from the first software stages of development process. This process is not always feasible, because it is not easy to change how developers work. The core opportunity for energy savings regarding computer monitors is to reduce the amount of energy used in active (on) mode. Reducing the amount of energy used in on mode is the largest energy-saving opportunity because computer monitors spend about 30% of the time in this mode. About 20% of the computer monitors in the market today meet the ENERGY STAR® Version 7.0 standards. The proposed regulations for mainstream computer monitors are slightly more stringent than the ENERGY STAR Version 7.0 specification and about 30% more stringent than ENERGY STAR Version 6.0. About 14% of current models would meet the proposed standards. Most monitors would need to reduce only their power consumption by 3–5 W to comply. This goal can be met by replacing components with efficient light-emitting diode lights, light-emitting diode drivers, and power supplies that are available in the market at prices comparable to the inefficient technologies.

Supportable HCI is a sub-field of **h**uman-**c**omputer **i**nteraction (HCI) that spotlights on the association amongst people and aptitude out of sight of sustainability [2]. Practical HCI had its beginning stage in 2007, when E. Blevis initially exhibited the possibility of Sustainable Interaction Design (SID). Supportability was measured a noteworthy paradigm for the plan of information, as vital in the outline procedure as criteria, for example, ease of use or strength.

Green and sustainable software engineering is the art of progressing green software engineering process therefore it is art of describing and promoting software products in a way [27]. So that the negative and positive impacts on sustainable development that result and or expected to result from the software product over it is whole life cycle are continuously assessed, documented and used for a further optimization of the software products [19].

According to [21], as identified different definitions sustainability, in general it considered from three different dimensions that are provided by the UN that is social, economic and environmental sustainability.

"Green IS and IT" represent to IS and IT products. The author describes what is Green IS and IT. Here, they tried to find out how the software is handled by an organization and how they maintain the emission and practices of disposal of IT spares without affecting the environment, whose objective is to avoid the pollution; Green Information System as inclusive of Green IT comprehensively has to be followed by the developers (people) and in software processes and technologies, and has to be maintained by person or group of person and public goals of program and prospective to influence on the sustainability of software company and communities giving threat for climate transform and other environmental aspects of dreadful conditions (**Figure 2**).

Figure 2.
Sustainability dimensions.

In a broad sense, computers are everywhere and consist of both specialized and generic systems. This report focuses on computers that constitute significant loads in buildings and specifically investigates energy-efficiency opportunities in five broad computer form factors: desktops, notebooks, small-scale servers, thin clients, and workstations. While the number of tablets in homes is increasing, the energy use of these products is relatively low, and the opportunity for savings is minimal due to existing battery charger regulations and market pressure to achieve high efficiency to enhance battery life. Therefore, this staff report does not include analysis on tablet computers. In homes, the most common form factors are notebooks and desktops. While there are more notebooks than desktops in California, the energy consumption of a desktop is more than double that of a notebook. This energy consumption increases when computer monitor energy use is included, which is necessary for functionality. **Table 1** shows estimates of home computer energy consumption with estimates ranging between 2.5 and 4.4% of all home electricity use, not accounting for computer monitor consumption.

Study	Representative Year	Computer Type	Number of Units (Millions, Scaled to CA[20])	Energy Use Per Unit (kWh/yr)	Total Energy Use (GWh/yr)	Percentage of Residential Electricity[21]
EIA MELS Analysis[22]	2011	Desktop	12.8	220	2,816	3.1%
		Notebook	20.6	60	1,236	1.4%
		Total	**33.4**	-	**4,052**	**4.4%**
CEA 2013 Residential Study[23]	2013	Desktop	11	186	2,046	2.2%
		Notebook	11.6	53	615	0.7%
		Total	**22.6**	-	**2661**	**2.9%**
ITI Comment[24]	2013	Desktop	9.6	187.3-296.4	1,800-2,800	2.0 - 3.1%
		Notebook	8.6	58.3-144.7	500-1,200	0.6 - 1.3%
		Total	**18.2**	-	**2,300-4,000**	**2.5 - 4.4%**

3.7 Criteria and metrics for sustainability

Criteria and measurements for supportability of a product items to spoke to in three categories [20] there are

1. Common quality and measurements.

2. Directly related criteria and measurements.

3. Indirectly related criteria measurements.

Based on the first-order effects of ICT supply, Second order effects of ICT use and Third order effects of Systemic effect of ICT respectively. Hardware obsolescence should be a genuine quality property of Green and Sustainable Software,

S. No	Product	Description
1.	Desktop computers	Generally paired with one or more computer monitors, displays, or televisions. Less commonly, the computer is integrated with a screen that is referred to as an "integrated Desktop." Desktop computers are generally responsible for the power management of these devices and can have power management responsibilities for accessories as well, such as printers.
2.	Notebook computers	A computer screen is integrated in the unit, and upgrades and configurability are generally limited. Although they can offer similar functionality to a desktop computer, they are somewhat constrained by space and power dissipation.
3.	Workstations, thin clients, and small-scale servers	A workstation is a task-oriented computer designed for abnormally constant and high workload and durability. On the opposite side of the spectrum, a thin client contains barebones interface hardware that may rely on separate equipment (generally a server or networked virtual machine) to provide full functionality, such as data storage and computational power. A small-scale server is a desktop computer configured to run as a server. While most modern desktops can be used as servers, small-scale servers generally have atypical hardware features and different operating systems than generic desktop computers.

Table 1.
Estimate of home energy consumption of computers.

which belongs to the directly related criterion and metrics model part. Indirectly related criterion and metrics for green and sustainable software address second and third order effects induced by software product [27, 28].

4. Green IT and Green ICT

We expect that computerized joining has aggravated the innovations of calculation and media communications to a degree that makes their division out of date in this unique situation. The term "Green IT" got comfortable after the production of a Gartner report in 2007 [22] and was later on joined by "Green Computing," "Green Software," "Green Software Engineering," and "Green Information Systems (IS)."

The fundamental distinction between Green in IT and Green by IT [21] is the pretended by the IT and the concentration of the greenness. The electricity consumption of computers, computer monitors, and signage displays varies greatly,

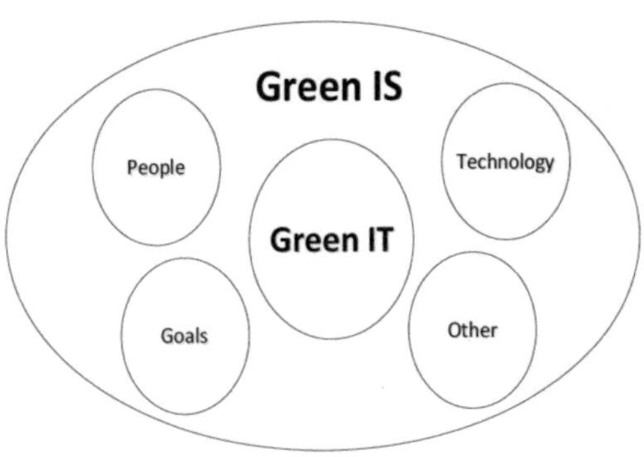

Figure 3.
Green IS and Green IT.

even within models of similar sizes and feature sets. To date, no federal or state regulations provide incentives for implementing cost-effective, readily available technologies to improve the performance of less efficient models. The five form factors considered in this report are desktops, notebooks, small-scale servers, workstations, and thin clients.

With help of the above factors we can find green in and by software and hardware (**Figure 3**). This combination creates the green software and hardware relationship which is Green IT.

4.1 Efficiency standards

The most substantial adjustments to the Energy Commission's draft staff report include dividing desktop computers into four categories with different efficiency levels rather than three categories with a smooth function, adjusting the levels of energy allowances for each category, and modifying the effective dates for the standards. These adjustments could be made while maintaining the magnitude of statewide energy savings. Staff's proposal is an effort to take international experience, stakeholder input, and data analytics and accomplish feasible and attainable energy savings for California in both long and short term. The proposed regulations are divided into three primary categories: desktops and thin clients, notebooks, and small-scale servers and workstations (**Figure 4**).

Category	Products Included
Desktops and thin-clients	Conventional desktops, thin clients, integrated desktops mobile gaming systems
Notebooks	Conventional notebooks, portable all-in-ones (AIOs), mobile thin clients, and two-in-one notebooks
Small-scale servers and workstations	Small-scale servers, high expandability computers (as defined), mobile workstations, rack-mounted workstations, and workstations.

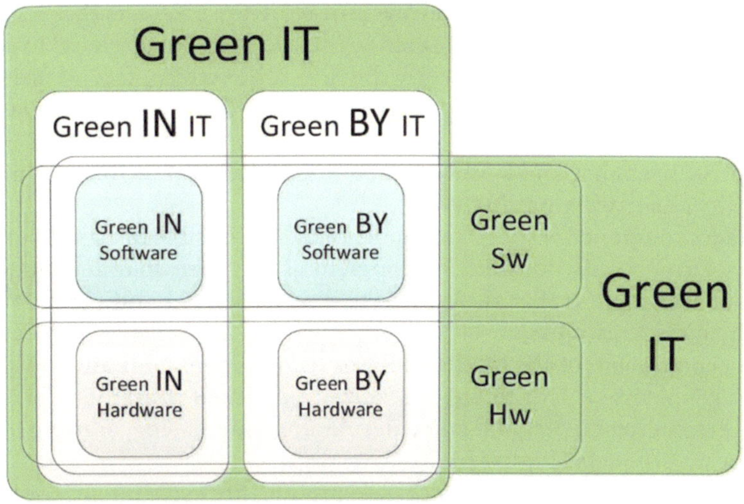

Figure 4.
Green software, green hardware and Green IT.

This procedure concentrates on what ought to be considered in each phase of the improvement routine collection, design, implementation, testing, deployment, and maintenance to create maintainable programming. This comes close to considers some of the classical water fall model methodology. In the requirement phase Marketing executive collect the requirement of the user, based on the software requirement condition the designer going to make a mockup design after getting design approval the developer or software engineer develop a coding, the test engineer going to test the software product whether it is working according to the software requirement specification finally the developing application implemented to the end user side.

5. Research methodology

This research aims at designing; implementing and evaluating new green-based software development life cycle as an alternate to the existing software development life cycle process with the aim of improving the Sustainability in Software Engineering Process.

New green-based SDLC modeling, measurement and evaluation related to Green database design have been identified as the three main approaches commonly used for evaluating the Green Software Engineering system. The primary aim of this research was concentrated on two phases. First, in the requirement phase in SDLC the green and sustainability is maintained with the help of cloud environment settings. Second, in the design phase the green and sustainability is maintained through the database design. The methodology of this study was software paper based data collection sheets, internet search engine and various tools have been studied [31].

5.1 GREENSOFT model

The GREENSOFT representation is a conceptual reference model for [20] "Green and Sustainable Software", Which has the aim to support software developers, administrators, and software users in creating, maintaining and using software in a more sustainable way, the model is shown in **Figure 1**, it comprises a holistic life cycle model for software products procedure models for different stakeholders and recommendations for actions that support stakeholders in developing, purchasing, supplying and using software in a green and sustainable manner.

The reference model contains a life cycle of software products that is in contrast to conventional life cycle software geared to life cycle thinking related to development, usage and end of life and identify the first-order effect, second order effect, and third order effect. The second part of the GREENSOFT model is called sustainability criteria and metrics it covers common metrics and criteria for the measurement of software quality and it allows taxonomy of parameters for evaluating the sustainability of software products.

The model component procedure model makes it possible to classify procedure models that cover acquisition and development of software, maintenance of IT systems and user support that also related to directly related, indirectly related and common quality metrics [35].

The last component of the Model contains recommendations and tools. These support stakeholders with different specialized skill levels in applying green or sustainable techniques in general when developing, purchasing, administrating, or using software products (**Figure 5**).

In the GREENSOFT model, the software product life cycle integrated ICT stimulated product life cycle that can also be credited with "from cradle-to-grave". It is intention to enable stakeholders to evaluate impacts on software development according to the three different levels of impacts [34].

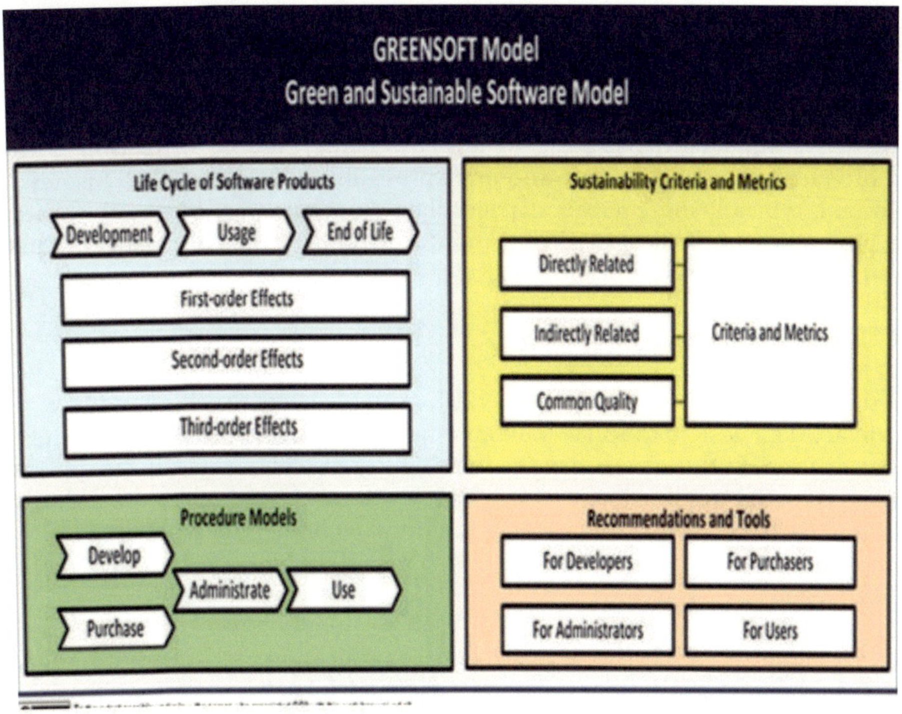

Figure 5.
Green and sustainable software model.

5.1.1 The development phase

The Development stage accounts for impacts [20] on software development that directly results from activities involved in software development as well as indirectly involved activities. Ecological impacts to be measured for example, Electrical Energy that is necessary to power the workstations of software developers and other employees electrical energy and natural resources that are necessary to operate the IT devices (example, Networking Devices, servers, and storages)(second order effects)Social impacts can be working conditions and payment of offshore employees (example developers, typesetters). Third order effects changes in organizations software development approach or life styles.

5.1.2 The distribution phase and the disposal phase

This phase [20] considered the things of software progress that distributing the software product that also includes the environmental impacts (example printed manuals, transportation used for travels, design used for the retail packing (e.g., plastic, biodegradable material or medium of data used for CD/DVD, and user memory Sticks) now a day's software product is offered to download a manuals at time considered the download size of the document [31].

5.1.3 The usage phase

It is impacts that result from deploying [20], maintaining and using of the software product that is nothing but software patches installation and giving training for employees to software usage which regarding in less energy consumption or just switch off the computer when they leave their workplace. To offer services by computer program requires the consumption of services offered by other servers like data base management systems or Enterprise Resource Planning (ERP). It also causes additional power consumption. To update of a software, the product size influences records transfer process in the IT based infrastructure which are necessary to deliver updates. It also basis further power and resource utilization to the new software to want up-to-date and more prevailing hardware [13]. This new hardware is typically more power efficient than older hardware but on the other hand production of the new hardware and the disposal of the old hardware causes resources and energy consumption [32].

5.1.4 The deactivation phase

If the data cannot be changed easily [20], example because it is stored in a proprietary data design, this may have an impact on economic sustainability of an organization. In this phase even the backing size of data format, if the domain transfers from one application platform to new domain platform. So the deactivation of old database format leads to economic and technical sustainability.

5.1.5 Discussion

The result of this study shows that software application was not environmental friendly in the software development life cycle model. Lack of sustainability is a major flaw in the existing green software model in the following phases such as Requirement, Design, Coding, Testing and Implementation. There are a lot of negative impacts such as high level energy usage, e-hazardous waste, ecological system, financial etc. In this research we proposed a new Green based Software Development Life Cycle in designing; implementing and evaluating phases for a sustainable development to overcome the above factors.

6. Procedure models

At the middle the designing procedure, the whole Life Cycle of the built programming items must be considered [20]. Develop, purchase, administrate, and use. In development of sustainability review study and preview study, process evaluation, sustainability presentation taken into consider about GREENSOFT model. In purchase of software product similar to the ENERGY STAR$^{(R)}$ that indicates whether a software product is energy efficient or meet certain sustainability requirement in the future. In administrate making software available by installing, configuring, and maintaining it. procedure model should implement a continuous improvement cycle that is energy utilization, should be checked regularly in order to improve these with appropriate measures apart from the data center operation includes networking, desktop PC, installed software resources utilization are need to be monitored in the administrative to procedure for users both the professionals and home users need to maintain the guidelines or checklist related to green and sustainable software [33].

6.1 Proposal and methods

Proposal and methods address collaborator with various roles [20] General roles review by the GREENSOFT model are: Planner, Customer, Organizer, and User. On the other hand, there may be more functional roles like Software Architect, Web Administrator, Requirements Engineer, or Application Developer. In assumption, these can be classified by the general roles mentioned before. Proposal can be best practice examples, utilization reports, guidelines, schedule, etc. Methods can be software, paper-based data collection sheets. There are plenty of tools available on the Internet. More knowledge base or professional Internet search engine would help to find easier.

7. Conclusion

The existing models are used to maintain green and sustainable software in the development and process evaluation. The GREENSOFT model has explained three order effects such as, the first-order effect focus on development to end, second order concentrate on sustainability area and last one focus on recommendations and tools for IT users in creating, maintaining a product for the sustainable development. Our approach is to develop Green IT Star model for green environment which focuses on five phases such as design, implementation, and requirement, coding and testing. The aim of this research is to help IT users in developing a green software and green hardware and improving the Sustainability in Software Engineering Process using Green IT Star model and has attained eco-friendly environment in design and requirement phase. The future research work is to achieve the Green software engineering in each Software Development Life Cycle phases on implementation, coding and testing phase.

Author details

Mohankumar Muthu[1*], K. Banuroopa[1] and S. Arunadevi[2]

1 Department of CS, CA and IT, Karpagam Academy of Higher Education, Coimbatore, India

2 Department of Science and Humanities, Kumaraguru College of Technology, Coimbatore, India

*Address all correspondence to: mohankumar07@gmail.com

References

[1] Du W, Pan SL, Zuo M. How to balance sustainability and profitability in technology organizations: An ambidextrous perspective. IEEE Transactions on Engineering Management. 2013;**60**(2):366-385

[2] Hilty LM, Aebischer B. ICT for sustainability: An emerging research field. In: Hilty LM, Aebischer B, editors. ICT Innovations for Sustainability. Springer Int. Publishing; 2015. pp. 3-36

[3] Sierszecki K, Mikkonen T, Steffens M, Fogdal T, Savolainen J. Green software: Greening what and how much? IEEE Software. 2014;**31**(3):64-68

[4] Amsel N, Ibrahim Z, Malik A, Tomlinson B. Toward sustainable software engineering: NIER track. In: 33rd International Conference on Software Engineering (ICSE). 2011. pp. 976-979

[5] Dick M, Naumann S. Enhancing software engineering processes towards sustainable software product design. In: Greve K, Cremers AB (eds) EnviroInfo 2010: Integration of Environmental Information in Europe. Proceedings of the 24th International Conference on Informatics for Environmental Protection; 6-8 Oct 2010; Cologne/ Bonn, Germany: Shaker; 2010

[6] IDC. Aid to Recovery: The Economic Impact of IT, Software, and the Microsoft Ecosystem on the Global Economy; 2009

[7] Koomey J. Growth in Data Center Electricity Use 2005 to 2010. Oakland, CA: Analytics. 1 Aug 2011. Available from: http://www.analyticspress.com/ datacenters.html

[8] Tate K. Sustainable Software Development: An Agile Perspective. Upper Saddle River, NJ: Addison-Wesley; 2006

[9] Murugesan S. Harnessing green IT: Principles and practices. IT Professional. 2008;**10**(1):24-33

[10] Tomlinson B. Greening Through IT. Cambridge, MA: MIT Press; 2010

[11] Kansal A, Zhao F, Liu J, Kothari N, Bhattacharya AA. Virtual machine power metering and provisioning. In: Proceedings of the 1st ACM Symposium on Cloud Computing. ACM; 2010. pp. 39-50

[12] Mingay S. Green IT: The New Industry Shock Wave. Gartner Inc; 2007

[13] Webb M et al. Smart 2020. Enabling the Low Carbon Economy in the Information Age. London: The Climate Group; 2008

[14] Capra E, Formenti G, Francalanci C, Gallazzi S. The impact of MIS software on it energy consumption. In: European Conference of Information Science 2010. 2010

[15] Shah A, Christian T, Patel CD, Bash C, Sharma RK. Assessing ICT's environmental impact. IEEE Computer. 2009;**42**(7):91-93

[16] Birol F. World Energy Outlook. International Energy Agency; 2010

[17] Elliot S, Binney D. Environmentally sustainable ICT: developing corporate capabilities and an industry-relevant is research agenda. In: Pacific Asia Conference on Information Systems; 4-7 July 2008; Suzhou, China; 2008

[18] Baer WS, Hassell S, Vollaard BA. Electricity Requirements for a Digital Society. Santa Monica, CA: RAND; 2002

[19] Naumann S, Dick M, Kern E, et al. The GREENSOFT model: A reference model for green and sustainable

software and its engineering. SUSCOM. 2011;**1**(4):294-304. DOI: 10. 1016/j. suscom.2011.06.004

[20] Naumann S, Dick M, Kern E, Johann T. The GREENSOFT model: A reference model for green and sustainable software and its engineering. Sustainable Computing: Informatics and Systems. 2011;**1**(4):294-304

[21] Calero C, Bertoa MF, Moraga MA. A systematic literature review for software sustainability measures. In: GREENS 2013: Second International Workshop on Green and Sustainable Software. 2013. pp. 46-53

[22] Gartner. Green IT: the new industry shockwave. In: Presentation at Symposium/ITXPO Conference. 2007

[23] Avaialble from: http://cbs.teriin.org/publication.php

[24] Kern E, Dick M, Naumann S, Guldner A, Johann T. Green software and green software engineering – definitions, measurements, and quality aspects. In: First International Conference on Information and Communication Technologies for Sustainability. 2013. pp. 87-94

[25] Kern E, Guldner A, Naumann S. Including software aspects in green IT: How to create awareness for green software issues. In: Kharchenko V, Kondratenko Y, Kacprzyk J, editors. Green IT Engineering: Social, Business and Industrial Applications. Studies in Systems, Decision and Control. Vol. 171. Cham: Springer; 2019

[26] Guldner A, Garling M, Morgen M, Naumann S, Kern E, Hilty LM. Energy consumption and hardware utilization of standard software: Methods and measurements for software sustainability. In: Otjacques B, Hitzelberger P, Naumann S, Wohlgemuth V, editors. From Science to Society. Progress in IS. Cham: Springer; 2018

[27] Madan S, Sukhdip SS, Bhatia R. Identification, selection and prioritization of sustainable and green software factors using fuzzy set theory (March 19, 2019). International Journal of Computational Intelligence & IoT. 2019;**2**(1):394-401. Available at SSRN: https://ssrn.com/abstract=3355396

[28] Jagroep E, van der Werf JM, Brinkkemper S, Blom L, van Vliet R. Extending software architecture views with an energy consumption perspective. Computing. 2016:1-1621

[29] Jagroep E et al. Awakening awareness on energy consumption in software engineering. In: 2017 IEEE/ACM 39th International Conference on Software Engineering: Software Engineering in Society Track (ICSE-SEIS). Buenos Aires; 2017. pp. 76-85

[30] Calero C, Piattini M. Puzzling out software sustainability. Sustainable Computing: Informatics and Systems. 2017;**16**:117-124, ISSN 2210-5379

[31] Nakagawa, Yumi E, Capilla R, Woods E, Kruchten P. Sustainability and longevity of systems and architectures. Journal of Systems and Software. 2018;**140**:1-2

[32] Venters CC, Capilla R, Betz S, Penzenstadler B, Crick T, Crouch S, et al. Software sustainability: Research and practice from a software architecture viewpoint. Journal of Systems and Software. 2018;**138**:174-188. ISSN 0164-1212

[33] Anthony BJ, Majid MA, Romli A.An empirical study on predictors of green sustainable software practices in Malaysian electronic industries. Journal of Information and Communication Technology. 2018;**17**(2):347-391

[34] Anthony BJ, Majid MA. Green IS for sustainable decision making in software management. Journal of Soft Computing and Decision Support Systems. 2016;**3**:20-34

[35] Seyff N, Betz S, Duboc L, Venters C, Becker C, Chitchyan R, et al. Tailoring Requirements Negotiation to Sustainability. In: 2018 IEEE 26th International Requirements Engineering Conference (RE). 2018. pp. 304-314

Chapter 6

International Cooperation for Smart and Sustainable Agriculture

Diana Dragomir, Mihai Dragomir, Daniel Acs and Sorin Popescu

Abstract

This chapter presents international best practices, realized within Europe, and focuses on cooperation for developing innovation support mechanisms and approaches in the area of smart agriculture. Specific situations are presented and analyzed in detailed regarding the requirements of smart agriculture and the possibilities to implement its percepts. As a consequence, solutions are proposed both in the technical and management domains to help speed up the transition from classical agriculture techniques to technology infused approaches, suitable for the current needs of this sector. Also, policy recommendations are developed based on the scientific findings in alignment with the evolution of the competitive pressures.

Keywords: smart agriculture, international cooperation, Danube region

1. Introduction and state of the art

Economic development sits at the crossroads of two megatrends that are developing fast and are ready to change the way human societies think and act about the future. One of these is sustainable development that attempts to change models, approaches, and cultures to balance the competitive impetuosity with the needs and limitations of the supporting ecosystems, while the other is the digital transformation (a.k.a. the smart revolution) which means fast, independent, and ubiquitous computers and electronic device processing large amounts of data continuously. Although these axes are very visible in manufacturing (e.g., Industry 4.0), automotive, consumer electronics, and even e-government, they are also present to a large extent in the field of agriculture and rural development, and they have an even more clear impact here because many areas of these sectors are a bit left behind in terms of development especially in developing countries. This is also the case in Central and Eastern Europe which came out of the communist period with an outdated agriculture that relied on mechanization and chemical products rather than biosciences and ecologically sound approaches. Moreover, the necessary process of restoring property rights further leads to de-evolution as small landowners had to gain technical and financial proficiency in order to rebecome competitive after a few decades.

This chapter deals with developments in smart agriculture cooperation in Romania and Slovakia, two countries that used to be part of the Eastern Block and faced similar but also specific challenges and which are now finding a new identity as part of the Danube macro-region coordinated and financed by the European Union (EU). The cross-cultural links among west and east along the Danube river are very good premise for establishing cooperation in the area of innovation support to help revitalize the agricultural sector in the 12 countries involved.

Scientifically speaking, smart agriculture is a trendy topic with significant developments being published in the last years. We will focus our next analysis on the situation in Europe, Romania, and Slovakia, addressing some important contributions both in the technical domain and in the economic one (**Table 1**).

Of course, this presentation is not exhaustive due to space limitations and a focus that does not include all the scientific disciplines connected with smart agriculture (e.g., chemistry, materials science, biotechnology, etc.). Its role is to provide an overview of the landscape that hosts the approaches described below in which the authors have been directly involved.

Scientific content	Type of contribution	Geographic scope	Source
Data mining study of 17,700 papers that shows the position of Europe as lagging in precision agriculture research and identifies a progression from topics related to crop management toward sustainability and sensorics	Literature review	Global/Europe, Italy	Pallottino et al. [1]
Investigation of technical approaches to machine learning applications in the areas of crop, livestock, water, and soil management, underlining their importance for future full-scale artificial intelligence deployment	Literature review	Not defined/global	Liakos et al. [2]
State-of-the-art study on the role of big data approaches for the development of smart farming, including closed vs. open access models	Literature review	Not defined/global	Wolfert et al. [3]
Development, testing, and performance review of an online cloud-based platform for small smart farm management	Practical achievement	Romania	Colezea et al. [4]
Case study on image processing of satellite photography for determining land destination and testing of the accuracy of the method	Practical achievement	Romania	Herbei et al. [5]
Solution building for a cyber-physical system that provides real-time monitoring and intervention in supervising potato cultivated fields	Theoretical study	Romania	Rad et al. [6]
Economic and environmental benefits of implementing precision techniques for the use of pesticides in crop management	Theoretical study	Hungary, Romania, EU	Takács-György et al. [7]
Creating dataset maps through data fusion in order to support the scenario-based policy interventions, with possible applications in agriculture	Practical achievement	Slovakia	Pazúr and Bolliger [8]
Algorithmic intercountry parallel investigation of the performances obtained by company processing agricultural products	Empirical study	Czech Republic, Slovakia	Čechura and Malá [9]
Mode of employment, results analysis, and improvement opportunities related to employing precision agriculture solutions	User survey	Five countries in the EU	Barnes et al. [10]

Table 1.
Comparative analysis of sustainable development scenarios.

2. Current situation of smart agriculture in Europe

The European Union and to a larger extent all the countries of Europe, as they have strong ties to the union, are searching for a pathway to competitiveness for a long time now. First came the Lisbon Strategy, then the Europe 2020 Strategy, and now the Future of Europe toward 2030 is being discussed. These fundamental guidelines helped maintain Europe on a strong development course in terms of economic growth, social inclusion, and competitiveness through troubling times like the 2008 financial crises and significant structural transformation like Industry 4.0. In all these documents, the issues of environmental accountability and efforts to protect the diversity of European ecosystems have been in the spotlight, constituting a signature trait of the union in the international arena. On the operational plane, sectoral strategies for agriculture and bioeconomy have been developed in the past years that include the concept of "smartness," thus fostering the appearance of smart and precision agriculture policies, funding instruments, technologies, solutions, and implementers. The support for this approach has led to the development of a competitive agricultural sector while at the same time ensuring the protection and safeguarding of the environment. This easily noticeable within the European Innovation Partnership "Agricultural Productivity and Sustainability" initiative acts as an innovation highway between EU's rural development programs and research and development programs and their associated stakeholders [11].

The main goals and directions of intervention of the EU bioeconomy strategy are summed up in **Figure 1**.

The three main axes are targeting sea and oceans, the replacement of fossil fuels and resources with bioresources (i.e., that can be grown), and the food and energy security of European citizens. Agriculture plays an important role as the source for many of the raw materials needed to implement these changes. Also, it is in its turn affected by the need to reduce the water footprint and the usage of fertilizers, while at the same increasing the yields and the quality of agricultural products. There are multiple ramifications to finding solutions relating to these issues, with smart and

Figure 1.
EU's approach to bioeconomy [12] (figure adapted by the authors).

precision agriculture dealing with some of them, while genetic engineering and circular economy are also forces to be reckoned with.

The future common agricultural policy (CAP) 2021–2027 has nine objectives that will reshape the sector within the European Union in the next 7 years (**Figure 2**):

Most of these priorities can benefit from the implementation of smart agriculture-based approaches in terms of sensors and precision guidance of agricultural equipment, Internet of things (IoT), and cloud solutions for infield interventions and ubiquitous computing and big data analytics for optimization and waste reduction in the production and processing stages.

Among the component structures of the Standing Committee on Agricultural Research at European level (SCAR), the strategic working group Agricultural Knowledge and Innovation Systems (AKIS) had been working for almost a decade on developing policy recommendations for supporting innovation frameworks in bioeconomy [14]. These guidelines will become part of the CAP strategic plans addressing the above-mentioned priorities and are expected to further boost the competitiveness of the agricultural sector in Europe.

In most Central and Eastern European countries, the implementation of smart agriculture is still developing, although some interesting solutions (e.g., anti-hail rockets launched based on computerized weather forecasts) are implemented and coexist with traditional farming methods. In our studied case, Slovakia can be considered a good practice for Romania, with important smart agriculture solutions being deployed on a considerable scale. This makes the domain ready for a massive influx of know-how, which can only come from a wide geographical area (e.g., the Danube region) and that will have significant impacts in the early stages of the digital transformation.

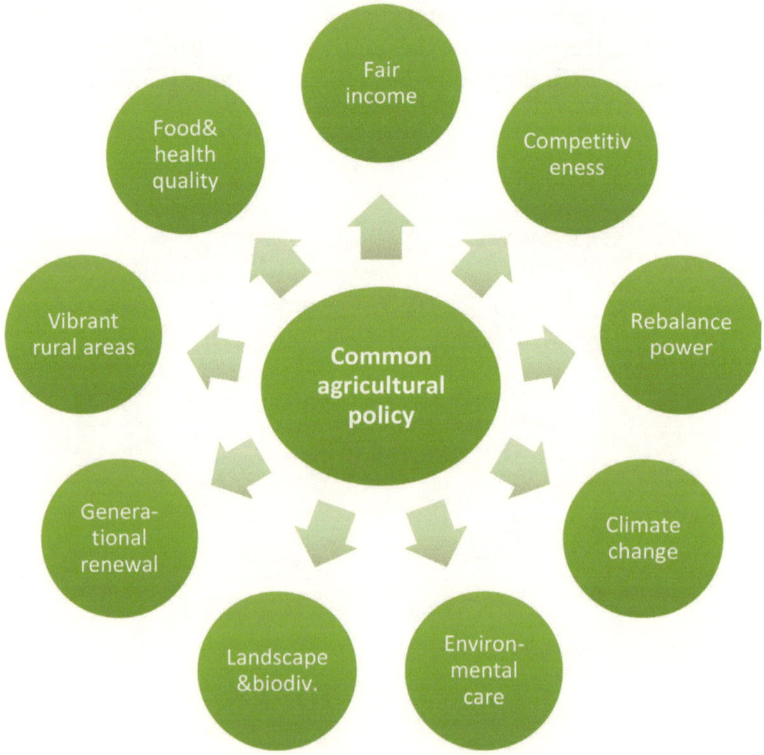

Figure 2.
EU's priorities for agriculture [13] (figure adapted by the authors).

Although other European countries are more advanced and have had for decades an industrialized agricultural sector, they are also facing important challenges in implementing smart approaches. As part of the effort to increase the trust of consumers, especially in the new bio- and eco-products as alternatives to the mass production of foodstuff, there is considerable contribution that can be made by the use of smart devices and software to process data and monitor agricultural production and product parameters. Mass implementation of such measures is desired by the customers and can be achieved faster in the new paradigm.

3. The Danube transfer centers network: a collaboration framework

The Danube transfer centers network (DTCN) has been growing strong since 2012 and has been an active participant in the field of technology transfer, innovation support, and transnational cooperation. It stems from a pilot initiative of the government of the state of Baden-Württemberg (federal state of Germany) implemented by Steinbeis Europa Zentrum (SEZ) and Steinbeis-Donau-Zentrum (SDZ) according the Steinbeis model in this domain, which is a success story on a national and European level. Starting with three pilot centers in Nitra, Novi Sad, and Cluj-Napoca, it has expanded through several cycles coordinated by SEZ to include centers in Bucharest (RO), Ruse (BG), Slavonski Hrast/Vukovar (HR), Maribor (SI), Pannon/Györ (HU), and Craiova (RO).

This very wide presence in the Danube region makes it a good choice for any stakeholders (companies on the one side and research institutions on the other side) to seek assistance in finding partners, solutions, or new project ideas to develop together. The network is predicated on the belief that transnational knowledge transfer is one of the main keys of sustainable development, economic growth, and social inclusion [15] and that the Danube region, due to its dimension and diversity, can be a good practice model for the entire European Union.

The network has developed an important online presence, and the platform www.dtcnetwork.eu hosts both presentation pages for the centers and links to instruments and tools for training, communication, and project management. The most important tool developed in-house by the Danube transfer center (DTC) Cluj-Napoca within the Interreg-DTP project "Made in Danube" is called Danube Transnational Innovation Cooperation (DTIC) and is full online system for partner matching and innovative project development from idea to results. It is available free of charge at this address, http://www.muri.utcluj.ro/tin-etool/index.php?page=login, and is operational for more than 1 year already (see **Figure 3**).

Due to the nature of the Danube region, specifically focused on the river and its related ecosystem, as well as due to the characteristics of the projects undertaken so far in common, the DTCN has developed a focus on eco-responsible innovation with preoccupations for bioeconomy, renewable energy, and international outreach toward the Eastern partnership countries and Western Balkan countries. Agriculture and food production are integral parts of this approach, and the need for smart agriculture solutions has become more noticeable over time, in conferences, bilateral talks, or DTIC platform searches.

DTC Cluj-Napoca, where three of the authors of this chapter are active, has a networked structure in itself and includes offices that activate in four universities in the city of Cluj-Napoca (the technical, the social and natural, the medical, and the agricultural ones), and one university (a comprehensive one) in the city of Sibiu. This creates multiple opportunities for interdisciplinary contacts among scientific disciplines and research areas, including modern agriculture and modern technology. The fourth author leads the Union of Slovak Clusters and is in close contact

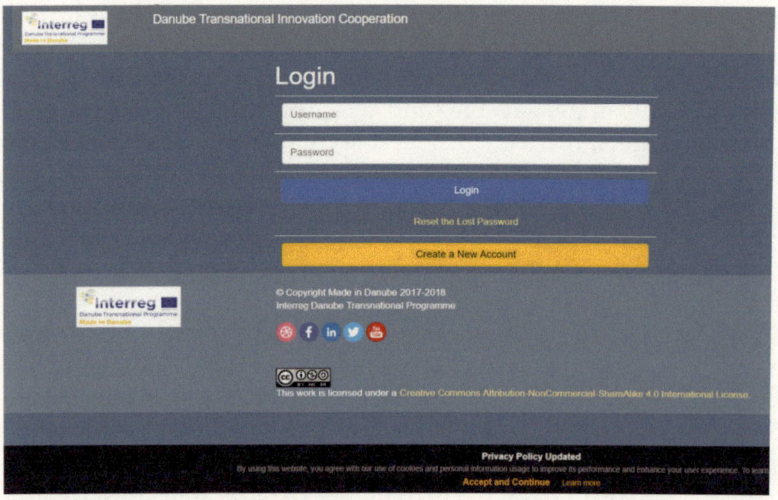

Figure 3.
Login interface of DTIC [16].

in the city of Nitra with the national agricultural university there which also hosts DTC Nitra, thus bringing into the current contribution an international perspective on the same topic.

Besides establishing innovation and technology transfer relations among their stakeholders, the DTCs in the network also undertake activities on a local scale or collaborate with each other on larger projects, usually with European or Danubian profile. Among the main activities on the network level, we can mention the following:

- Cooperating in European research, development, and innovation projects funded by EU's framework programs

- Cooperating in transnational framework and policy development projects financed by EU's interregional programs

- Participating in associations, clusters, and networks with thematic and sectoral characteristics in other countries than the host one

- Maintaining a consistent image, a common or aligned web, and social media presence to project the scope of the collaboration

- Organization or participation in common to relevant events in the macro-region: the annual forum, brokerage events, conference, workshops, etc.

- Exchange of good practices, experts, and know-how in the form of visits, trainings, bilateral projects, and development of specific competence centers in line with smart specialization strategies and needs

All these elements contribute to developing a stronger network that is also oriented toward territorial and content-wise expansion to match the true development potential of the Danube region, which is judged by all those involved to be considerable and with a long-time halo. In this respect, there are also many challenges to face in the present and the future. Some of those identified so far are presented below:

- Distinct cultural approaches to international cooperation and the inclusion in the region of EU and non-EU countries, with different relations to the so-called European project and obvious language differences
- Large economic, infrastructure, and development disparities which bring about significant differences in terms of needs, interests, and preoccupations regarding research and innovation of all stakeholders
- A “digital divide” with high penetration and low speeds in the West and low penetration and high speeds in the East
- Differences in legal and accounting systems that make collaboration difficult and time-consuming, especially in bureaucratic grants and projects

4. The “Made in Danube” project approach

The “Made in Danube” project (full name “transnational cooperation to transform knowledge into marketable products and services for the Danubian sustainable society of tomorrow”) is currently being implemented (time frame 2017–2019) with financing from the Interreg-Danube transnational Program (project code DTP1-1-072-1.1) and involves seven of the DTCs active in the Danube region, being focalized on supporting the development of bioeconomy in this area of the continent through innovation and technology transfer policies and instruments. The outputs and results of the project reflect the contributions of the partners toward researching and supporting this emerging economic sector, with demonstrators proposed on three directions, each in a different country [17]:

- Smart and innovative precision farming—implemented in Nitra, Slovakia
- Competence Center in Wood Sector—implemented in Vukovar, Croatia
- Biofuels—implemented in Novi Sad, Serbia

The project has a total of six workpackages (**Figure 4**), with the Technical University of Cluj-Napoca being in charge of the one aimed at developing new tools in instruments and the Union of Slovak Cluster contributing to the pilot local action plan implementation in Nitra that deals with innovative precision farming, in partnerhsip with the Slovak University of Agriculture [18]. Thus, these two entities have expanded their collaboration on the topic of smart agriculture that started within the DTCN, with the DTIC platform (developed by TUCN) playing an important role in the activities carried out as part of the Slovakian pilot initiative.

The project aims to bring together all relevant actors working in bioeconomy in the macro-region (companies—especially small- and medium-sized ones, professional networks, universities, research institutes, nongovernmental organizations, public authorities at local and national levels, experts, and the general public) and to perform the scanning and cross-referencing of strategies, policies, and other programmatic documents to contribute in the future the better alignment of interests and initiatives. The web platform, the direct technology transfer instruments, and the training materials represent the main vehicle through which changes are being designed in line with state-of-the-art concepts and approaches related to innovation, before being deployed through the localized demonstrators and proposed as solutions on regional level to decisionmakers.

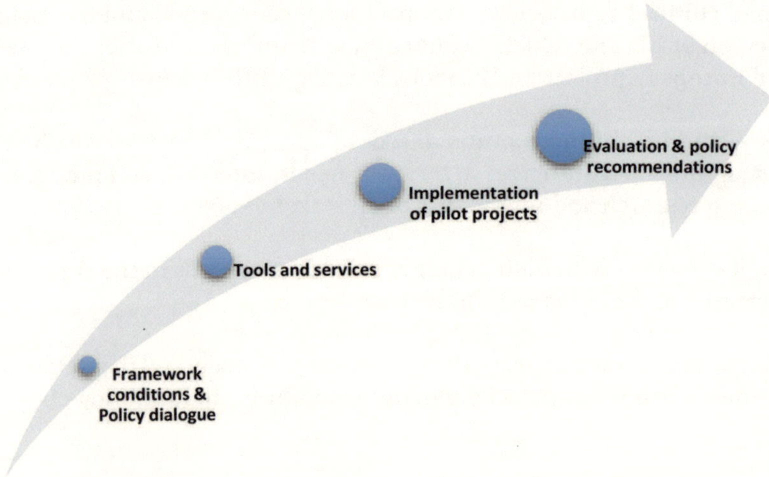

Figure 4.
Logical structure of "Made in Danube" activities [18].

5. Some results in smart agriculture

This section presents two contributions in which international cooperation, established with the help of the previously mentioned instruments, has contributed to better and faster smart solutions to be implemented in the farming sector.

The first project deals with developing a new type of tractor for a Romanian manufacturing company based on agricultural expertise from Slovakia and Internet of things expertise from Hungary. The main development themes set by the company for the new product lines dealt with the following topics:

- Following in a manned and semi-unmanned fashion precise working trajectories to ensure better yields
- A climate smart device with a lower carbon footprint than current tractors existing on the market
- Real-time data collection of information regarding soil and weather conditions to adapt the operations to the work conditions
- Ability to work under difficult conditions with ease

The product development process has followed a combined model, using elements of stage-gate and quality function deployment, with the documentation for production approaching finalization. **Figure 5** presents a summary of the deployment of requirements to product characteristics, identifying along the way the main subassemblies in relation to the agreed upon technical specification.

The chassis design and the fitting of the engine and powertrain were undertaken in Romania, the mobile sensor configuration and connectivity was designed in Hungary, and the adaptation of the device to agricultural best practices was achieved in Slovakia. This collaboration has been undertaken based on a cooperation agreement, with the three involved organizations sharing costs and risks in the same percentage as future sales of the product, should it be successful in the promising market [19]. The intellectual property rights have been mapped out from the beginning, and common authorship patents will be filed for the innovative elements.

The second project deals with a multinational team of scientist and experts from the Danube region and also from the UK and the USA involved in developing a total preventive maintenance model for automated screw conveyors used in grain silos. The product combines a mechanical structure with automation and sensorics that permit to start and stop according to the quantity and flow of the product that it has to transport. It is used for loading and unloading operations, in relation with trucks and human operators and, to a larger extent, in internal transport operations of the silos in order to achieve the rotation of stock and treatment operations upon the grains and for optimizing loads and usage of the storage spaces that form the facility.

The equipment operates mostly automatic, with manual control possible in case of override situations. Both the mechanical and the automation components require

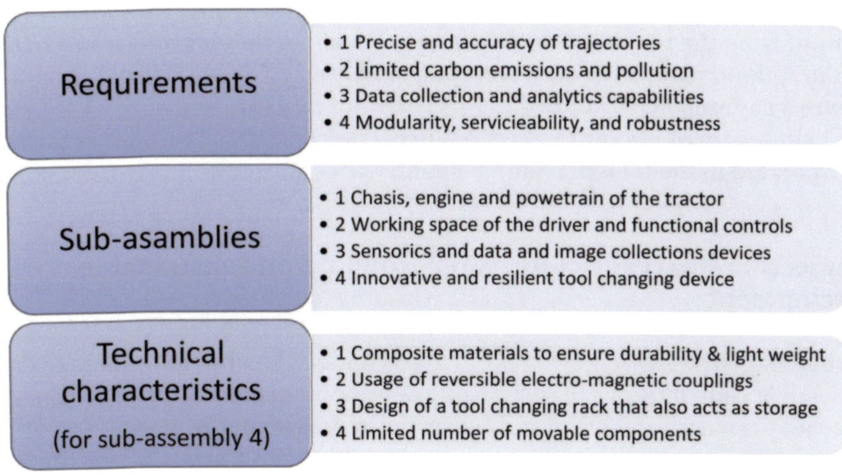

Figure 5.
Establishing the main smart tractor components.

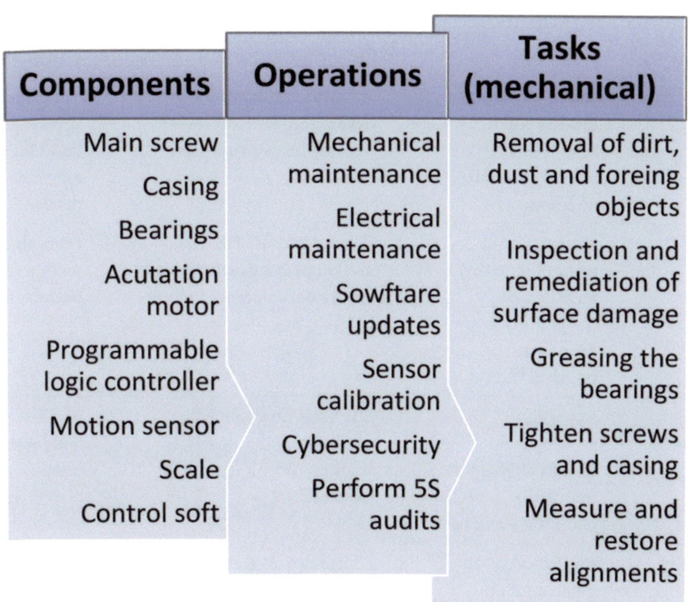

Figure 6.
Total preventive maintenance for an automated screw conveyor.

considerable preventive maintenance in order to remain in an operating condition inside the environment of the silos, which is characterized by high amounts of dust, large variation of temperature with the outside weather, and possible blockages when grains enter the conveyor components and get lodged. The total preventive maintenance (TPM) model is applied with the help of two maintenance teams. A maintenance program is applied through software, based on a risk management algorithm that determines the components most prone to breakdown, taking into account complexity and history of operation. The software permits the grouping and scheduling of operations and the recording of maintenance dates, including tasks performed and situations encountered in the field, as well as the generation of material list for supplementing the consumables and materials stock (measuring devices, bearings, controllers, plugs, grease, paint, etc.).

By implementing the smart conveyor instead of the classical version and by applying the total maintenance program on a silo with six conveyors (one for input and output from the building and five internal ones for moving and transporting the grains), the savings have been recorded to be over 6.5% of the total revenue per year, with a payback period of less than 3 years but an estimated active life (with proper maintenance) of cca. 15 years. **Figure 6** presents the main elements that can be incorporated in the total preventive maintenance strategy.

6. Connecting smart agriculture, bioeconomy, and sustainable development

In this chapter, we will present the analysis undertaken by the authors to determine possible contributions to sustainable development coming from modern approaches to smart agriculture and bioeconomy, based on the description of best practices in the Danube region. For this purpose, the well-known 17 sustainable development goals (SDG) proposed by the United Nations [20] are used as a reference frame. A qualitative description is presented in **Table 2**, and a quantitative approach, based on the binational expertise from Slovakia and Romania captured

UN SDG [20]	No change	Smart agriculture	Full bioeconomy
Goal 1: No poverty	Current economic models still have high levels of poverty and scarcity of resources	Subsistence farming in developing economies will become more productive	Considerable impact across agriculture, manufacturing, recycling, and energy production
Goal 2: Zero hunger	Hunger is still present in many locations on the globe due to economic and political factors	Better yields will drive market prices down and will ensure better access to food	More abundant and more nutritious food will become available
Goal 3: Good health and well-being	Existing development models cannot keep up with population growth and increasing needs	The need for chemical fertilizers and industrial farming techniques will be reduced	Besides agriculture and food production, cosmetics and pharmaceuticals can benefit from bioeconomy
Goal 4: Quality education	Education is not given enough priority especially in terms of accessibility	Some improvements can come about but are rather limited	Considerable research and innovation are needed, and education will have to change to meet demands

UN SDG [20]	No change	Smart agriculture	Full bioeconomy
Goal 5: Gender equality	The gender imbalance is changing slowly	There will be no additional contribution in this respect	More improvements should happen consistently
Goal 6: Clean water and sanitation	Industry, agriculture, and transportation contribute to massive water quality problems	Agricultural runoff and resource usage will significantly decrease, improving water quality	All economic sectors will produce less waste, pollution, and runoff that affect water resources
Goal 7: Affordable and clean energy	Fossil fuels play a major role in meeting the energy demands for economic development	Precision farming will generate more biomass and can also correlate with wind-/wave-based energy	There will be extra energy conservation by reusing processed materials and improving biomass yields
Goal 8: Decent work and economic growth	Economic growth is slow and worker salaries tend to not be sufficient for decent living	People in the agricultural sector can move onto services or other advanced sectors	Bioeconomy requires renewed industries for horizontal and vertical integration
Goal 9: Industry, innovation, and infrastructure	Current trends can be maintained, with a broad and diverse focus of innovation efforts	Agriculture and supporting domains will gain an innovation boost to implement precision farms	Implementing bioeconomy will require considerable R&D effort in fundamental and applied sciences
Goal 10: Reduced inequality	Income inequality among countries and inside countries is widespread	Some minor improvement will ensue due to the commoditization of food	Considerable entrepreneurship and innovation opportunities are to be expected
Goal 11: Sustainable cities and communities	There are considerable difficulties in ensuring a balanced social and economic development	Rural communities will become smaller and more technologically advanced, while cities will receive better products	Additional connections within and among communities will be necessary to ensure new value chains
Goal 12: Responsible consumption and production	Current consumption patterns are unbalanced and affecting ecological balances	An important component will be improved/reduced—food and bioresources (e.g., fodder, raw materials)	In a bioeconomy model, most materials and components are recycled and reused
Goal 13: Climate action	There are severe climate consequences to the current economic and industrial development	The contribution of agriculture to climate change will be reduced through process optimization	Bioeconomy is one of the keys to addressing climate change, but other policy and technology measures are also needed
Goal 14: Life below water	The health of the world oceans is severely affected by pollution and biodiversity reduction	Smart agriculture will reduce the ecological footprint upon ocean waters	Agriculture and manufacturing have less impact, but transport and overfishing remain
Goal 15: Life on land	The economic situation is still difficult for many countries and everyday life is affected for their people	Improvements will take place in terms of food quality and availability, but job opportunities in agriculture will be reduced	Quality of life under bioeconomy is improved in all relevant areas: nutrition, health, future prospects, etc.

UN SDG [20]	No change	Smart agriculture	Full bioeconomy
Goal 16: Peace and justice strong institutions	The UN and democratic states are working toward this goal, but many changes are still needed	There is a limited connection between better food and work options and institutional development	Changing economic balances and lifting countries out of poverty should improve peace prospects
Goal 17: Partnerships to achieve the goals	There are common efforts and cooperation initiatives being implemented	The impact on international cooperation is limited due to the constrains of the sector	New companies, regions and countries will develop in new directions, improving relations

Table 2.
Comparative analysis of sustainable development scenarios.

Criteria \ New Concepts	1 Stay the course - No change	2 Smart and precision farming	3 Full scale circular bio-economy	Importance
2 GOAL 1: No Poverty	-	+	++	5,0%
3 GOAL 2: Zero Hunger	-	+	++	10,0%
4 GOAL 3: Good Health and Well-being	O	+	++	5,0%
5 GOAL 4: Quality Education	O	O	+	10,0%
6 GOAL 5: Gender Equality	O	O	O	5,0%
7 GOAL 6: Clean Water and Sanitation	--	+	++	5,0%
8 GOAL 7: Affordable and Clean Energy	-	O	+	5,0%
9 GOAL 8: Decent Work and Economic Growth	O	+	++	5,0%
10 GOAL 9: Industry, Innovation and Infrastructure	+	+	++	5,0%
11 GOAL 10: Reduced Inequality	-	O	+	5,0%
12 GOAL 11: Sustainable Cities and Communities	-	O	++	5,0%
13 GOAL 12: Responsible Consumption and Production	--	+	++	5,0%
14 GOAL 13: Climate Action	--	O	+	10,0%
15 GOAL 14: Life Below Water	--	-	+	5,0%
16 GOAL 15: Life on Land	O	+	++	5,0%
17 GOAL 16: Peace and Justice Strong Institutions	+	+	++	5,0%
18 GOAL 17: Partnerships to achieve the Goal	+	+	++	5,0%
Positive Effects	3	10	16	
Negative Effects	9	1	0	
Neutral Effects	5	6	1	
Net Effect	-6	9	16	
Positive Priorization	5,0%	18,3%	71,7%	
Negative Priorization	-35,0%	-1,7%	0,0%	
Net Effect	-30,0%	16,7%	71,7%	

Figure 7.
Weighted comparison of scenarios—Pugh's method.

with the help of the Pugh method implemented in the Qualica QFD software, is displayed in **Figure 7**. Three scenarios are analyzed in parallel for a time frame of 11 years (2019–2030), with respect to the contribution in realizing the SGD: staying the current course with no significant change (basal scenario), implementation of smart agriculture/precision farming (realistic scenario), and full-scale implementation of bioeconomy including biotechnologies and bio-based industries (optimistic scenario).

This analysis served as basis for the implementation of a selection methodology known as Pugh's method that assigns weights to criteria (in our case the SDG) and ranks the alternatives based on neutral, negative, and positive effects. All the goals received a 5% importance rating, except three considered priorities that received 10% (hunger elimination, quality education, and climate action). The results are presented below:

As it can be noticed, the current course has a slow positive progression, the implementation of smart agriculture contributes significantly, but a full bioeconomy approach on a world-wide scale would be much more effective.

7. Conclusions

There is a significant potential for developing and implementing smart agriculture solutions in the Danube region, both in terms of policies and scientific contributions, and the elements presented in this chapter constitute building blocks of a proper ecosystem for this. Agriculture has historically been a strong sector for both Romania and Slovakia, and there are national policies as well as private initiatives attempting to recapture this competitive advantage in the form of smart devices, technologies, or projects. There is a good and diverse capability for developing this domain (strong IT sector, developed universities in the technical and life sciences areas, fast Internet, and good penetration of technology in rural areas), and we believe cooperation frameworks, like the Danube Transfer Centers Network and the Interreg-Danube projects, can contribute to transforming this capability through proper policies and instruments into concrete results. This is even more timely in the present with increasing discussion about possible food crises in the future, as well as an increasing focus on finding biological and ecological solutions for supporting a circular and sustainable industry, like growing fuels, construction materials, and ingredients specific to the pharmaceuticals and cosmetics.

The international and transnational dimensions of cooperation in this sector come to complement the economic driving axis, because smaller countries that are cooperating in macro-regional (i.e., Danube region) or supranational (i.e., the European Union) contexts have improved chances of being competitive and developing fast in the current setting of a globalized economy. As proposed in the chapter, smart farming is only the first step in implementing a full-scale bioeconomy approach and should be undertaken soon to help change the status quo.

Author details

Diana Dragomir[1*], Mihai Dragomir[1], Daniel Acs[2] and Sorin Popescu[1]

1 Technical University of Cluj-Napoca, Romania

2 Union of Slovak Clusters, Slovakia

*Address all correspondence to: diana.dragomir@muri.utcluj.ro

References

[1] Pallottino F, Biocca M, Nardi P, Figorilli S, Menesatti P, Costa C. Science mapping approach to analyze the research evolution on precision agriculture: World, EU and Italian situation. Precision Agriculture. 2018;**19**:1011-1026

[2] Liakos KG, Busato P, Moshou D, Pearson S, Bochtis D. Machine learning in agriculture: A review. Sensors. 2018;**18**:2674

[3] Wolfert S, Ge L, Verdouw C, Bogaardt MJ. Big data in smart farming—A review. Agricultural Systems. 2017;**153**:69-80

[4] Colezea M, Musat G, Pop F, Negru C, Dumitrascu A, Mocanu M. CLUeFARM: Integrated web-service platform for smart farms. Computers and Electronics in Agriculture. 2018;**154**:134-154

[5] Herbei MV, Popescu CA, Bertici R, Smuleac A, Popescu G. Processing and use of satellite images in order to extract useful information in precision agriculture. Bulletin UASVM series Agriculture. 2016;**73**(2):238-246

[6] Rad CR, Hancu O, Takacs IA, Olteanu G. Smart monitoring of potato crop: A cyber-physical system. Agriculture and Agricultural Science Procedia. 2015;**6**:73-79

[7] Takács-György K, Rahoveanu T, Magdalena M, Takács I. Sustainable new agricultural technology—Economic aspects of precision crop protection. Procedia Economics and Finance. 2014;**8**:729-736

[8] Pazúr R, Bolliger J. Enhanced land use datasets and future scenarios of land change for Slovakia. Data in Brief. 2017;**14**:483-488

[9] Čechura L, Malá Z. Technology and efficiency comparison of Czech and Slovak processing companies. Procedia Economics and Finance. 2014;**12**:93-102

[10] Barnes AP, Soto I, Eory V, Beck B, Balafoutis A, Sánchez B, et al. Exploring the adoption of precision agricultural technologies: A cross regional study of EU farmers. Land Use Policy. 2019;**80**:163-174

[11] EIP-AGRI. European Innovation Partnership 'Agricultural Productivity and Sustainability' [Online]. 2019. Available from: https://ec.europa.eu/eip/agriculture/en/european-innovation-partnership-agricultural [Accessed: March 1, 2019]

[12] European Commission. Review of the 2012 European Bioeconomy Strategy; 2017

[13] European Commission. Future of the Common Agricultural Policy [Online]. 2018. Available from: https://ec.europa.eu/info/food-farming-fisheries/key-policies/common-agricultural-policy/future-cap_en [Accessed: February 10, 2019]

[14] SWG SCAR-AKIS. Mission and Aims [Online]. Available from: https://scar-europe.org/index.php/akis-mission-and-aims [Accessed: March 1, 2019]

[15] Danube Transfer Centres Network. Home [Online]. 2015. Available from: http://www.dtcnetwork.eu/ [Accessed: February 25, 2019]

[16] Made in Danube Consortium, TUCN. Login [Online]. 2018. Available from: http://www.muri.utcluj.ro/tin-etool/index.php?page=login [Accessed: February 25, 2019]

[17] Made in Danube Consortium. Home [Online]. 2017. Available from: http://www.interreg-danube.eu/approved-projects/made-in-danube [Accessed: February 22, 2019]

[18] Made in Danube Consortium. Made in Danube Application Form. 2016

[19] Knowledge Sourcing Intelligence LLP. Global Smart Tractor Market—Forecasts from 2018 to 2023; 2018

[20] United Nations. About the Sustainable Development Goals [Online]. 2018. Available from: https://www.un.org/sustainabledevelopment/sustainable-development-goals/ [Accessed: April 4, 2019]

Section 4

Case Studies

Chapter 7

Green Building Rating Systems as Sustainability Assessment Tools: Case Study Analysis

Mady Mohamed

Abstract

Building performance and occupants' comfort lie at the core of building design targets. Principles of green architecture and building physics are not given enough thought and consideration. In the best cases, some thought is given to such factors but without a scientific methodology, which takes into consideration appropriate climatic data and appropriate assessment tools. Most importantly, the interference of the environmentalist in architecture projects comes usually very late in the design processes. Facing these facts has driven most countries to adopt official strategies and policies to deal with building's performance. The rating systems are among these initiatives. The author of this chapter adapts a detailed methodology to aid the integration of the principles of the green architecture in the early stages of design using rating systems. The Leadership in Energy and Environmental Design (LEED) 1 that was developed in the USA by the U.S. Green Building Council (USGBC) for Core and Shell has been employed as the main design target. This chapter presents a brief about the world green initiatives and discusses the results of applying the methodology of integrating the green architectural principles at the early stages of design processes—through precedent analysis.

Keywords: rating system, LEED, sustainability targets, computer simulation, post occupancy evaluation

1. Introduction

More than half of the world's population lives in cities; in 2050 the people living in urban areas are expected to increase up to 70% [1]. Cities are the major reasons of pollution; it produces 60% of carbon dioxide and greenhouse gas emissions, through using energy generations, industry, vehicles, and biomass use. Therefore, now climate change is challenging cities to reduce their impacts and adjust to changing condition [2]. Therefore, the increasing demand towards sustainability is pushing toward rapid changes in policies, laws, and regulations around the world regarding products and processes to encourage more sustainable projects [3]. Also, sustainability solves the local issues of communities in innovative progress, for implementing sustainability is different for every community, but they share common goals for a healthy environment, smart growth, and human well-being [4]. Consideration to sustainability principles in building industry is vital for natural environment and human being. Adopting passive strategies and measures that respond to and achieve the responsive design lies directly under the responsibility of architectural designers [5]. Green architecture principles and science are usually not given enough thought and consideration. Factors such site characteristics, climate, and orientation, environmental design of the building, and choice of building materials are being neglected in most cases. In the best cases, some thought is given to such factors but without a scientific methodology that takes into consideration using the appropriate climatic data and the appropriate assessment tools. Consequently, buildings often have a poor indoor environment quality which in turn affects human comfort, health, and efficiency [6, 7]. Most importantly, the interference of the environmentalist in architecture projects comes usually very late in the design processes. Consequently, buildings often have a poor indoor environment quality which in turn affects human comfort, health, and efficiency. Most importantly, the interference of the environmentalist in architecture projects comes usually very late in the design processes. The integration of these green principles in the field at the early stages of the design processes lies at the core of the current research. However, to get the best benefits of these strategies and measures, detailed target identification must be set. Adopting these concepts has driven most countries to adopt official strategies and policies in order to insure appropriate building designs.

2. Sustainability and green buildings initiatives

The application of sustainability is carried out by different stakeholders including academic initiatives, government initiatives, other sector initiatives, in addition to private sector initiatives. Moreover, these initiatives vary in its nature and way of application; some of them are building standards and codes, framework and programs, in addition to rating systems [8].

The standard is a set of guidelines and criteria to assure the quality of the products. Standards related to building industry are created by organizations such as the International Standards Organization (ISO), which defines and develops worldwide standards that frequently become law or form the basis of industry norms. ISO defines a standard as "a document, established by consensus, approved by a recognized body that provides for common and repeated use as rules, guidelines, or characteristics for activities or their results." There are other institutions such as the American National Standards Institute (ANSI), the American Society for Testing and Materials (ASTM), or the American Society of Heating, Refrigeration, and Air-Conditioning Engineers (ASHRAE) [8].

Green codes could be classified into two types: prescriptive and performance, with outcome-based as a third option. A prescriptive path is a fast, definitive, and conventional approach to code compliance. It provides tables to quantify certain levels of strictness for materials and equipment. Performance-based codes are designed to achieve certain results, rather than meeting prescribed requirements for individual building elements. Outcome-based codes establish a consumption target for energy, water, etc. One example of the green codes is the International Green Construction Code (IgCC) that provides a comprehensive set of requirements in order to reduce the harmful effects of buildings on the natural environment [8].

Programs and frameworks are database that provide datasets relating to most aspects of sustainability. Examples of such programs and frameworks are the RFCS, CPI, CPDP, and UNEP-SBCI. The reference framework of sustainable cities (RFSC) encourages sustainability and integrated urban development aligned with Europe 2020 guidelines and objectives [9]. The city prosperity initiative (CPI) measures sustainability at the urban level to allow local and central governments to use data [10]. The climate positive development program (CPDP) addresses the challenges of rapid urbanization and climate change [11]. In addition, there is the United Nations Environment Program-Sustainable Building and Climate Initiative (UNEP-SBCI), which is a partnership of major public and private sector stakeholders in the building sector, working to promote sustainable building policies and practices worldwide [12]. The Passivhaus standard was developed in Germany in the early 1990s, and the first dwellings to be completed to the Passivhaus standard were constructed in Darmstadt in 1991 [13].

Rating systems assess the environmental impacts of buildings, constructions, infrastructure, urban-scale project, and community projects. The rating systems designed to assist projects to be more sustainable by providing frameworks with a set of criteria's that cover several aspects of a project's environmental impact [14]. Rating systems utilize the key performance indicators (KPI) to assure high quality of sustainability applications [14]. KPI are employed for building designers and decision-makers to measure the socioeconomic and environmental impacts on environment, infrastructure, waste system, regulations, pollutions, citizen's access to services, and more [15]. The significance of the sustainable design increased in the 1990s. The Building Research Establishment's Environmental Assessment Method (BREEAM) was the first green building rating system in the UK that addressed the required KPIs for better environmental performance of buildings. In 2000, the U.S. Green Building Council (USGBC) developed another rating system, which is the Leadership in Energy and Environmental Design (LEED). Others also responded to the growing interest and demand for sustainable design including additional rating systems that most of them were influenced by these early programs but are tailored to their own context with specific priorities. Other trails for rating systems intended to address broader issues of sustainability or evolving concepts such as social aspects, net zero energy, and living and restorative building concepts. It is estimated that there are nearly 600 green product certifications in the world with nearly 100 in use in the USA, and the numbers continue to grow [16]. Many other rating systems became a great evidence of adapting the sustainability principles in building industry [17, 18]. The rating system is based on four major components [14].

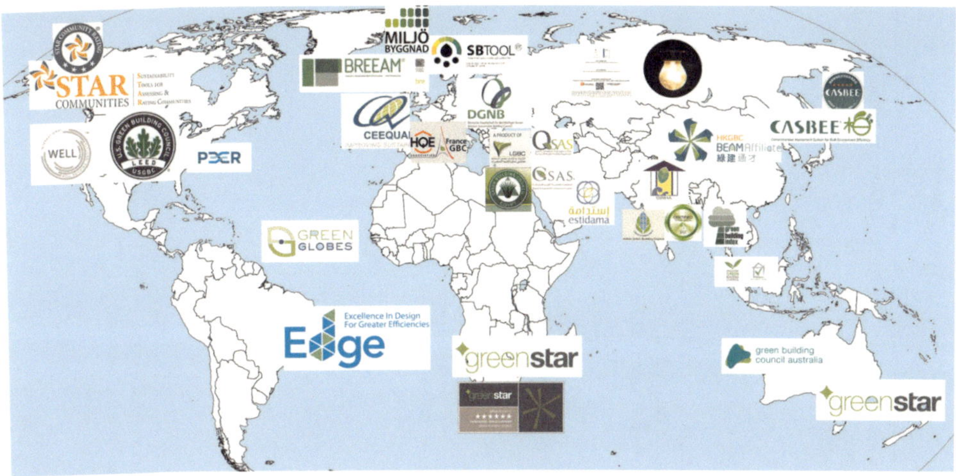

Figure 1.
Common rating assessment systems around the world (by the author).

Tool	Year	Country	Targets	Main categories for buildings	Certification levels	Development basis	Notes and aim
Building Research Establishment Environmental Assessment Method (BREEAM)	1990	UK/International	• Buildings • Interiors • Infrastructure • Master planning projects	Energy and water use Internal environment (health and well-being) Pollution, transport Materials, waste ecology, and Management processes	Pass, Good, Very good, excellent	Original	It is the world's leading sustainability assessment tool to recognize and reflect the value in higher-performing assets across the built environment lifecycle, from new construction to in-use and refurbishment [24] (https://www.breeam.com/)
Leadership in Energy and Environmental Design (LEED)	2000	USA/International	• Buildings • Interiors • Neighborhood development • Cities and communities	Sustainable Sites Water Efficiency Energy and Atmosphere Materials and Resources Indoor Environmental Quality (IEQ) Innovation in Design Regional Priority	Certified, Silver, Gold, Platinum	Original	LEED is available for virtually all building, community, and home project types. LEED provides a framework to create healthy, highly efficient, and cost-saving green buildings. LEED certification is a globally recognized symbol of sustainability achievement [25] (https://new.usgbc.org/leed)
Indian Green Building Council Rating Systems (IGBC Rating Systems)	2001	India	• Buildings • Interiors • Residential Societies • Cities and Communities • Villages • Health and Well-being Rating	Sustainable Architecture and Design, Site Selection, and Planning Water Conservation Energy Efficiency Building Materials and Resources, IEQInnovation and Development	Certified (Best Practices), Silver (Outstanding Performance), Gold (National Excellence), Platinum (Global Leadership)	LEED	All the IGBC rating systems are voluntary, consensus-based, market-driven building programs. The rating systems are a perfect blend of ancient architectural practices and modern technological innovations. The ratings systems are applicable to all five climatic zones of the country [26] (https://igbc.in/igbc/)
Comprehensive Assessment System for Built Environment Efficiency (CASBEE)	2001	Japan	• Buildings • Interiors • Heat Island • Urban Development • Cities	Energy efficiency Resource efficiency Local environment Indoor environment	_S _A _B+ _B _C [14]	Original	CASBEE was developed by a research committee established in 2001 through the collaboration of academia, industry, and national and local governments, which established the Japan Sustainable Building Consortium (JSBC) under

Tool	Year	Country	Targets	Main categories for buildings	Certification levels	Development basis	Notes and aim
			• Health checklist				the auspice of the Ministry of Land, Infrastructure, Transport and Tourism (MLIT) [27] [28] (http://www.ibec.or.jp/CASBEE/english/)
Green Globe International Standard for Sustainability	2002	International	• Accommodation and hospitality • Transport and tour operators • Conference venues • Meeting planners • Management and public relations	Energy Indoor Environment Site Water Resources Emissions Project/Environmental Management	Certified, Gold, Platinum	BREEAM	Green Globe provides certification, training and education, and marketing services in 83 countries worldwide for the sustainable operations and management of travel and tourism companies and their related supplier businesses [29] (https://greenglobe.com/)
The Green Star rating system (Green Star)	2002	Australia	• Communities • Buildings • Interiors • Operational Performance	Management processes IEQ Energy, Transport Water, Materials Land Use Ecology Emission, Innovation	Best Practice, Australian Excellence, World Leadership	BREEAM, LEED®	The Green Star rating system assesses the sustainable design, construction and operation of buildings, fit-outs, and communities [30] (https://new.gbca.org.au/green-star/rating-system/)
Performance Excellence in Electricity Renewal (PEER)	2003	US/International	• Power system performance • Electricity delivery systems	Reliability and resiliency Operations, Management and safety Energy efficiency and environment, Grid services Innovation and Exemplary Performance, Regional Priority	Certified, Silver, Gold, Platinum	LEED	PEER is the nation's first rating system that measures and improves power system performance and electricity delivery systems. Developed in a collaboration between the GBCI (Green Business Certification Inc.) and Bob Galvin, formerly of Motorola [31] (http://www.gbci.org/press-kit-peer)

Tool	Year	Country	Targets	Main categories for buildings	Certification levels	Development basis	Notes and aim
BCA Green Mark	2005	Singapore	• Buildings • Interiors • Districts • Infrastructure	Energy efficiency Water efficiency Environmental protection IEQ, and Other green and innovative features that contribute to better building performance	Certified, Gold, Gold Plus, Platinum	Undisclosed	The BCA Green Mark Scheme aims to drive Singapore's construction industry toward more environment-friendly buildings. It is intended to promote sustainability in the built environment and raise environmental awareness among developers, designers, and builders when they start project conceptualization and design, as well as during construction [32] (https://www.bca.gov.sg/green_mark/)
STAR Community Rating System	2007	USA	• Cities and Communities	Built Environment, Climate and Energy, Economy and Jobs, Education, Arts, and Community, Equity and Empowerment, Health and Safety, Natural Systems Innovation and Process	CERTIFIED: 3-STAR Community 4-STAR Community 5-STAR Community	LEED	Assess sustainability and measure progress to enhance the quality of life and human well-being [33] (http://www.starcommunities.org/)
Global Sustainability Assessment System (GSAS)	2007	Qatar	• Buildings • Districts • Infrastructures	Energy Water Indoor Environment Cultural and Economic Value Site Urban Connectivity Material Management and Operation	Urban Connectivity, Site, Energy, Water, Material, IEQ, Cultural and Economic Value, Management and Operations	LEED, BREEAM, Green Globe, CEPAS, CASBEE, and SBTool [34]	Create a sustainable environment that reduces the ecological impact and classify the social and cultural needs and the environment of the region [35] (https://www.gord.qa/gsas-trust)

Tool	Year	Country	Targets	Main categories for buildings	Certification levels	Development basis	Notes and aim
Green Star Tools	2007	South Africa	• Buildings • Interior fit-outs • Precincts	Management IEQ, Energy Transport Water, Materials Land Use and Ecology Emissions Innovation	4 stars, Best Practice; 5 stars, South African Excellence; 6 stars, World Leadership	Australian Green Star which is based on BREEAM and LEED	An internationally recognized and trusted mark of quality for the design, construction, and operation of buildings, interior fit-outs. and precincts [36] (https://gbcsa.org.za/certify/green-star-sa/)
DGNB Global Benchmark for Sustainability	2007	Germany	• Buildings • Interiors • Urban districts	Environmental Quality, Economical Quality, Sociocultural and Functional Quality, Technical Quality, Process Quality, Site Quality	Bronze, Silver, Gold, and Platinum	Original	The DGNB System provides an objective description and assessment of the sustainability of buildings and urban districts. Quality is assessed comprehensively over the entire life cycle of the building [37] (https://www.dgnb-system.de/en/system/certification_system/index.php)
Sustainable Buildings Tool (SBTooL)	2009	Lithuania	• Buildings	Site Regeneration and Development, Energy and Resource Consumption, Environmental Loadings, IEQ, Service Quality, Social, Cultural and Perceptual Aspects, Cost and Economic Aspects	Best Practice, Good Practice, Minimum Practice, Negative [38]	Original	SBTool is a generic framework for rating the sustainable performance of buildings and projects. It may also be thought of as a toolkit that assists local organizations to develop local SBTool rating systems [39] (http://www.iisbe.org/sbmethod)
Green Pyramid Rating System (GPRS)	2009	Egypt	• Buildings	Sustainable Sites Water Efficiency Energy and Atmosphere Materials and Resources Indoor Environmental	Certified, Silver Pyramid, Golden Pyramid,	LEED	Establishment of the Egyptian Green Building Council: It is to provide a mechanism to encourage building investors to adopt BEECs as well as other sections of existing

Tool	Year	Country	Targets	Main categories for buildings	Certification levels	Development basis	Notes and aim
				Quality Innovation in Design Regional Priority	Green Pyramid, [40]		codes that satisfy both energy efficiency and environmental conservation by focusing on new construction [41] (http://egypt-gbc.org/ratings.html)
Green Star NZ	2009	New Zealand	• Buildings • Interiors • Communities	Management processes IEQ, Energy Transport Water, Materials Land Use Ecology Emission Innovation	Good Practice, Best Practice, NZ Practice, World Excellence	BREEAM	Green Star is a tool to support stakeholders in the property and construction sectors to design, construct, and operate projects in a more sustainable, efficient, and productive way [42] (https://www.nzgbc.org.nz/GreenStar)
Building Environmental Assessment Method (HK BEAM Plus)	2009	Hong Kong	• Buildings • Interiors • Neighborhood	Site aspects Material aspects Water use Energy use Indoor environmental quality Innovations and additions	Bronze, Silver, Gold, Platinum	BREEAM	BEAM Plus assessment is to offer independent assessments of building sustainability performance. BEAM Plus certification is a proven path for creating safer, healthier, more comfortable, more functional, and more energy-efficient buildings [43] (http://greenbuilding.hkgbc.org.hk/)
GREENSL® Rating System for Built Environment (GreenSL)	2010	Sri Lanka	• Buildings	Management Awareness Sustainable Sites Energy and Atmosphere Materials and Resources Indoor Environmental Quality, Process Innovation, and Design Social and Cultural	Certified, Silver, Gold, Platinum	Undisclosed	A Green Environmental Rating System applicable to Sri Lanka has been formulated as a "home-grown system" with all norms acceptable to leading rating systems [44] (http://srilankagbc.org/rating.php#)

Tool	Year	Country	Targets	Main categories for buildings	Certification levels	Development basis	Notes and aim
Green Building Index (GBI)	2010	Malaysia	• Buildings including Historical Buildings • Interiors • Township	Sustainable Site Planning and Management Water Efficiency Energy and Atmosphere Materials and Resources IEQ Innovation in Design	Certified, Silver, Gold, Platinum	LEED	GBI is to promote sustainability in the built environment and raise awareness among developers, architects, engineers, planners, designers, contractors, and the public about environmental issues and our responsibility to the future generations [45] (http://new.greenbuildingindex.org/)
Green Rating for Integrated Habitat Assessment (GRIHA)	2010	India	• Buildings • Large Development • Cities	On-site Sufficiency; Water, Energy, Solid Waste Management Development Quality; Site Planning, Energy, Water, and wastewater management, Transport, Solid Waste Management, Socioeconomic	1 Star, 2 Stars, 3 Stars, 4 Stars, 5 Stars	Undisclosed	The GRIHA rating system consists of 31 criteria categorized under various sections such as Site Planning, Construction Management, Occupant Comfort and Well-being, Sustainable Building Materials, Performance Monitoring and Validation, and Innovation [46] (http://www.grihaindia.org/griha-rating)
Pearl Building Rating System (PBRS)	2010	Abu Dhabi	• Buildings	Integrated Development Process Natural Systems Livable Communities Precious Water Resourceful Energy Stewarding Materials Innovating Practice	1 Pearl, 2 Pearls, 3 Pearls, 4 Pearls, 5 Pearls	Undisclosed	The aim of the Pearl Building Rating System is to promote the development of sustainable buildings and improve quality of life. The PBRS encourages water, energy and waste minimization, and local material use and aims to improve supply chains for sustainable and recycled materials and products [47] (http://www3.cec.org/islandora-gb/en/islandora/object/greenbuilding%3A101)

Tool	Year	Country	Targets	Main categories for buildings	Certification levels	Development basis	Notes and aim
Miljöbyggnad (MB) “Environmental Building—system for sustainable building certification”	2010	Sweden	• Buildings	Energy, Materials, IEQ	Gold, Silver, Bronze	LEED and BREEAM	The MB system is based on the Swedish Building Regulations (BBR), which govern the entire country. It is relatively simple and includes only 15 items measured for certification [48] (http://insight.gbig.org/green-building-in-sweden-sgbc-miljobyggnad/ https://www.sgbc.se/certifiering/miljobyggnad/)
STO NOSTROY 2.35.4–2011 “Rating system for evaluation sustainability of residential and public buildings” **GOST R 54964–2012** “Environmental requirements for real estate”	2011/ 2012	Russia	• Buildings • Real estate	Quality of architecture, IEQ, Quality of sanitary protection and waste management, Operation, Training water management, Energy efficiency, Economic efficiency	Undisclosed	LEED, BREEAM, DGNB, and HQE	Buildings and civil construction. Rating system for evaluation sustainability of residential and public buildings and real estate. It defines the principles, categories, evaluation criteria, sustainability indicators of habitat, as well as weighting for ratings for buildings [21] (http://zvt.abok.ru/articles/47/Green_Building_Market_Situation_in_Russia)
Excellence in Design for Greater Efficiencies (EDGE)	2012	• World Bank Group • “Internationally”	• Buildings	Energy Water Materials	Pass/Fail [49]	Original	The EDGE application helps to determine the most cost-effective options for designing green within a local climate context. EDGE can be used for buildings of all vintages, including new construction, existing buildings, and major retrofits [50] (https://www.edgebuildings.com/marketing/edge)

Tool	Year	Country	Targets	Main categories for buildings	Certification levels	Development basis	Notes and aim
ARZ BRS Green Building Rating System	2012	• Lebanon	• Buildings	Energy Performance Thermal Energy Electrical Energy Building Envelope Materials, IEQ Water Conservation Operations and Management	Gold Silver Bronze Certified Registered projects	Edge	The ARZ Building Rating System is designed to measure the extent to which existing commercial buildings in Lebanon are healthy, comfortable places for working, and consuming the right amount of energy and water, while having a low impact upon the natural environment [51] (http://arzrating.com/)
Miljöbyggnad MB —"Environmental Building - system for sustainable building certification	2010	Sweden	• Buildings	Energy, Materials, and IEQ	Gold, Silver, or Bronze	LEED and BREEAM	The MB system are based on the Swedish Building Regulations (BBR), which govern the entire country. It is relatively simple, includes only 15 items measured for certification [48] (http://insight.gbig.org/green-building-in-sweden-sgbc-miljobyggnad/ https://www.sgbc.se/certifiering/miljobyggnad/)
High Quality Environmental (**HQE**) standard	2013	France/International	• Buildings • Urban Projects	Energy, Environment, Health, Comfort [23]	Pass, Good, Very good, Excellent, Exceptional	Original	HQE™ is the French certification awarded to building construction and management as well as urban planning projects. HQE™ promotes best practices and sustainable quality in building projects and offers expert guidance throughout the lifetime of the project [52] (https://www.behqe.com/)

Tool	Year	Country	Targets	Main categories for buildings	Certification levels	Development basis	Notes and aim
The WELL Building Standard® (**Well**)	2015	USA	• Health and Wellness of Buildings' occupants	Air, Water, Nourishment, Light, Fitness, Comfort, mind	Silver, Gold, Platinum	Original	It is a performance-based system for measuring, certifying, and monitoring features of the built environment that impact human health and well-being, through air, water, nourishment, light, fitness, comfort, and mind [53] (https://www.usgbc.org/articles/what-well)
Civil Engineering Environmental Quality (**CEEQUAL**) Assessment and Award Scheme	2015	UK/International	• Civil engineering • Infrastructure • Landscaping • Public realm projects	Project/Contract Strategy, Project or Contract Management, People and Communities, Land use and Landscape, The Historic Environment, Ecology and Biodiversity, Water Environment (fresh & marine), Physical Resources Use and Management, Transport [54]	Pass, Good, Very Good, Excellent	Original	CEEQUAL is the evidence-based sustainability assessment, rating, and awards scheme that challenges projects to deliver better outcomes in infrastructure sustainability, developed by the Building Research Establishment BRE, UK [55] (http://www.ceequal.com/)

Table 1.
Summary of the common rating assessment systems around the world (by the author after [8, 13, 19–23]).

- Categories: These form a specific set of items relating to the environmental performance considered during the assessment.
- Scoring system: This is a performance measurement system that cumulates the number of possible points or credits that can be earned by achieving a given level of performance in several analyzed aspects.
- Weighting system: This represents the relevance assigned to each specific category within the overall scoring system.
- Output: This aims at showing, in a direct and comprehensive manner, the results of the environmental performance obtained during the scoring phase.

Figure 1 and **Table 1** present the most common green rating systems all over the world chronologically. **Table 1** summarizes the most important features of those rating systems, in terms of year of establishment, coverage, main categories for building rating, level of certifications, its development base, and main aim with the main link of the source.

3. Case study analysis

In previous researches [56–58], the author of this chapter had set a detailed methodology to aid the integration of the principles of the sustainability in the early stages of design (**Figure 2**). The outputs of these researches have been employed in several real-life building projects on the regional level. The current research

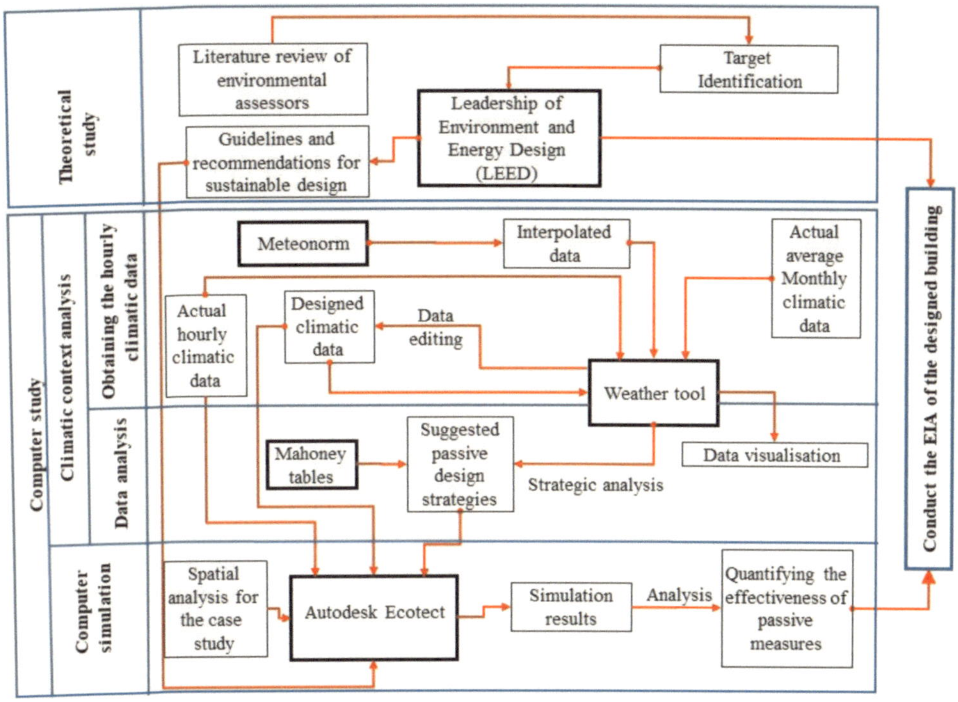

Figure 2.
Proposed detailed methodology to integrate the environmental assessment in the architectural design process [56].

presents one project as a case study analysis. The adopted methodology employs the environmental assessor "Leadership in Energy and Environmental Design" to measure the compatibility of the design with principles of sustainability. Also environmental software (Autodesk Ecotect, HTB2, and Weather Tool in addition to environmental tools such as psychometric chart, Mahoney tables, and Stereograph diagram and Solar Tool) have been used in order to analyze the context and quantify the effectiveness of proposed passive strategies and measures. By such, design proposals in the early stages of design (i.e., design concept, orientation of buildings, using passive strategies and techniques, facade designs and projections, colors of the buildings, opening size and design, etc.) could be quantified. LEED has 110 credits which cover all the different disciplines in building design and construction. However, the current application focuses on the related credits to the early stages of design which lie directly under the architect responsibility and can affect the total performance of the building.

4. Target identification

The adopted methodology employs the Leadership in Energy and Environment Design (LEED) 1 that was developed in USA by the U.S. Green Building Council for new construction as one of the most known environmental assessors in the market nowadays. The LEED tool aims to provide building stakeholders with a "report card" that indicates the health, efficiency, and comfort of the buildings. LEED recognizes the unique nature of the design and construction of ASHRAE Advanced Energy Design Guide [59] and addresses the specific needs of building spaces and occupant's health issues [60]. LEED is flexible to apply to all project types including healthcare facilities, schools, homes, and even the entire neighborhoods. LEED for Core and Shell can be used for projects where the developer controls the design and construction of the entire Core and Shell base building (e.g., mechanical, electrical, plumbing, and fire protection systems) but has no control over the design and construction of the tenant fit-out. Projects could include a commercial or medical office building, retail center, warehouse, or lab facility. It is designed to be complementary to LEED for commercial interiors and LEED for Retail: Commercial Interiors.

The allocation of points between credits is based on the potential environmental impacts and human benefits of each credit with respect to a set of impact categories. The impacts are defined as the environmental or human effect of the design, construction, operation, and maintenance of the building, such as greenhouse gas emissions, fossil fuel use, toxins and carcinogens, air and water pollutants, and indoor environmental conditions. A combination of approaches, including energy modeling, life cycle assessment, and transportation analysis, is used to quantify each type of impact. The resulting allocation of points among credits is called credit weighting [61]. These credit weightings are shown in **Figure 3**. LEED V4 are awarded according to the following scale in **Table 2**.

This work aimed at achieving the LEED Rating system (Core and Shell). Most of the LEED issues could be quantified by analyzing the design input data, while other issues such as Indoor Environmental Quality needs a quantification tool to be assessed. This methodology employs thermal comfort and energy efficiency as environmental design targets. The effectiveness of the proposed measures is determined according to its ability to passively achieve thermal comfort by using minimum amount of energy possible. This helps the designer to recognize successful LEED strategies and measurements for achieving credit category goals.

This work had set the guidelines for the architectural and engineering design of the GREENEDGE building based on analyzing the macroclimate for Cairo city and

LEED New Construction V4	
Category	Total Credits
Integrative Process	1
Location and Transportation	16
Sustainable Sites	10
Water Efficiency	11
Energy & Atmosphere	33
Materials & Resources	13
Indoor EnvironmentalQuality	16
Innovative in Design	6
Regional Priority	4
TOTAL	110

Figure 3.
The credits weighting of the environmental categories of the LEED, [7].

LEED ratings	LEED v3
Certified	40–49 points
Silver	50–59 points
Gold	60–79 points
Platinum	80+ points

Table 2.
Certification scale of LEED [18].

the microclimate data for the GREENEDGE site. This is done through using a specific scientific computer-based methodology developed by the author of the chapter through his research [6, 7, 56–58, 62–67] that mainly depends on a number of environmental design computer-based tools and especially the comprehensive environmental analysis and simulation tools. These tools are:

- The analysis sustainable building design software (Autodesk Ecotect)
- Climatic analysis software (Weather Tool)
- Solar analysis software (Solar Tool)
- Mahoney tables
- Shadowing analysis (Stereograph diagram)
- Synthesizing hourly climatic data (Meteonorm)

The use of computer software allow the visualization of the unseen environmental attributes in a three-dimensional interface, allowing by such comprehensive understanding of the issues involved in the assessment process.

5. Project understanding and location

The New Cairo Business Hub (GREENEDGE) is located at plot 84, First sector, New Cairo City Center, that is directly overlooking the southern 90 road right beside BNP Paribas Headquarters (**Figure 4**). The building is designed to be a class (A) office building with total plot area of 33,000 m2 of office spaces for banks and multinational companies at one of the most developed business districts in Egypt with all required amenities and facilities at place and surrounded by Egypt's biggest banks, headquarters, as well as notable multinationals.

5.1 Basic project information

Project name: THE GREENEDGE.
Land area: 7123 Sqm.
Footprint %: 25%.
No. of floors: Three basements + G + six typical floors.
Owner: Katamia for office Buildings—KOP.
Project developer: Redcon Real Estate Development.
Green Architecture and LEED Consultant: The author of the current chapter.

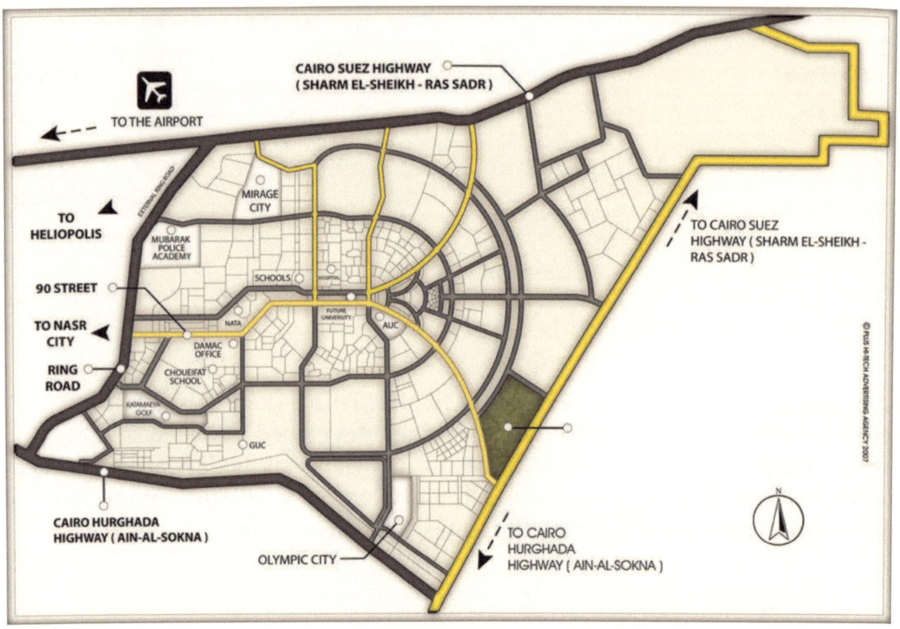

Figure 4.
New Cairo location and the location of GREENEDGE building, after Google maps [68] and new Cairo City Council [69].

6. Macroclimate analysis

Cairo's climate is a desert climate, which remains mostly dry and arid year-round. The hot weather in Cairo means that the humidity can rise at times, particularly during winter (December to February). At this time precipitation is more likely, and temperatures drop to 13–19°C. Cairo weather in the summertime (May to August) sees temperatures of 45–47°C. The Cairo International Airport weather Station was chosen to most represent the location of new Cairo. The hourly climatic data file generated by the USDOE was used in this report. On analyzing the hourly climatic data using Weather Tool, Cairo climate is classified as an arid climate where precipitation rarely occurs. Cairo has a hot desert climate (Köppen climate classification: BWh). The climate is generally dry. The temperatures are hot or very hot in summer days and warm or mild in winter days, but warm in summer nights and cool in winter nights. The temperature varies greatly, especially in summer; it ranges from 7°C at night to 40°C during the day. While the winter temperature does not fluctuate as wildly, it can be as low as 0°C at night and as high as 18°C during the day. Cairo receives less than 25 mm of precipitation annually in most areas and almost never rains in summer. Air temperatures are being outside the comfort zone most of the year. Only during 4 months (March, April, September, and October), a good percentage of the total hours is found to be located in the comfort zone. The prevailing wind is coming from the north to northwest most of the year with average air temperature, while hot wind comes from the west-south direction during specific times of the year. Prevailing wind are coming from the north to northwest most of the year with average air temperature, while hot wind comes from the west-south direction during specific times of the year. Rainfall is rare in Cairo and does not exceed 25 mm/the whole year.

Passive solar heating, thermal mass effect, night purge ventilation, natural ventilation, direct evaporative cooling, and indirect evaporative cooling to enhance the environmental performance of the GREENEDGE in Cairo were tested using Weather Tool. The analysis revealed that while thermal mass and night purge ventilation can enhance the thermal performance during the whole year, almost only natural ventilation can enhance significantly the environmental performance of the building during the summer season. While indirect evaporative cooling can enhance the thermal performance slightly during the summer time, passive solar heating can also contribute to the thermal enhancement during winter time. Using Mahoney table, it revealed that it is essential to deal with such climate to use the following strategies:

- Compact plans with interior courtyards
- Dual-targeting buildings that allow air circulation intermittent
- Small, 15 to 25% of the surface of the walls
- Openings in the north and south walls
- Construction heavy for strong thermal inertia for walls and roofs; jet lag more than 8 hours

7. Results and discussion

7.1 Existing design analysis

In this section, the original design of GREENEDGE building (**Figure 5**) will be explained, highlighting the problems, constrains, and potentials.

The GREENEDGE building in its base case was exposed to high incident solar radiation especially on its west and south facades that receive solar radiation every single day of the year with no any internal open spaces such as courtyards. This would affect negatively the building performance. Shaded open spaces are very preferable in the hot dry zones. They can reduce the daytime air and radiant temperatures inside the occupied space. The courtyard helps in maintaining cooled indoor temperatures. It provides a private internal open space that is visually and

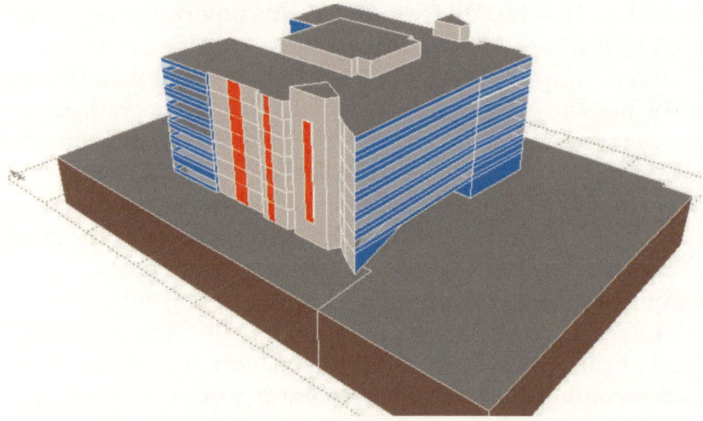

Figure 5.
The simulated GREENEDGE building as it is, done by the author after Autodesk Ecotect.

acoustically separated from the outside environment. The base case material for all the windows was single glazed that is not appropriate for such climate particularly for the west-south facades and high intensity of solar radiation.

7.2 Sustainable design enhancements

To deal with the current situation, several traditional and contemporary ideas have been adopted. The recommended ideas and solution could be classified under the recommended passive strategies that were raised from the climatic analysis using Weather Tool and Mahoney tables. This could be listed below.

7.2.1 Vegetation around the building

Maximizing the amount of vegetation inside and outside buildings affects positively the thermal performance of buildings. This could result in shading of the external surfaces of the building, shading the opened spaces, reducing and filtering the dust in the air, and elevating the humidity level [70]. However, vegetation in such climatic conditions could be expensive because of the limitation in the water supply for irrigation and by turn could be against the green architecture principles (**Figure 6**). Specific types of trees and irrigation technology should be selected to best suit the climatic context.

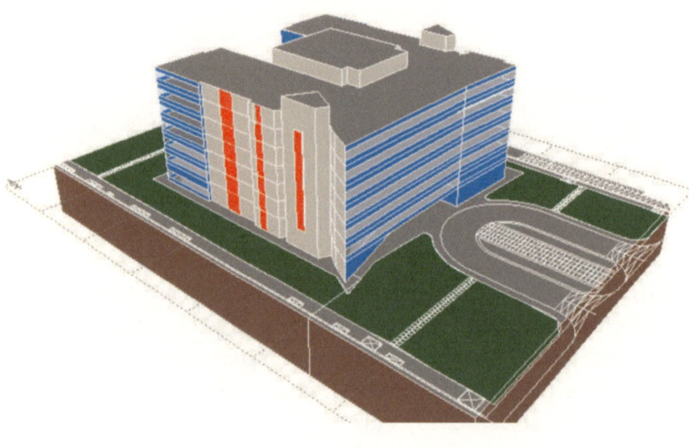

Figure 6.
Vegetation around the building (done by the author after Autodesk Ecotect).

Grass area has been avoided since it needs potable water for sprinkler irrigation system. According to the WHO guidelines for the use of treated wastewater for irrigation, gray water could not be used for adjacent area for man activity [71, 72], also because gray water can affect negatively the sprinkler heads. Moreover, high-efficiency drip irrigation systems can be 95% efficient, compared with 60–70% for sprinkler or spray irrigation systems [73]. Also, the use of native or adapted vegetation on the project site can assist project teams with earning more credits regarding sustainable sits.

7.2.2 Compact plans with interior courtyards that allow air circulation

A recommended northern courtyard with link between the courtyard and the backyard at the south orientation has been modified to the design. This can affect positively the thermal performance of the building. This link could be positioned at the first floors “called Takhtabush in vernacular architecture.” This could be achieved by replacing the curtain glazing in this area to contemporary electronic Mashrabia (**Figure** 7).

This ensures a steady flow of air by convection [74]. Since the backyard is larger at the south orientation, and thus less shaded than the courtyard, air heats up more than in the courtyard. The heated air rising in the backyard draws cool air from the courtyard through the Takhtabush, creating a steady cool breeze.

7.2.3 Openings in the north and south walls, the exposed side of the human height of the wind, and interior wall openings

Window height and details have been modified to be in two parts with different heights. Those of the north direction must be the same in height with the human being. Opposite ones must be in a higher position to enable the required cross-ventilation. This will give the occupants the controllability of opening the upper or the lower parts according to the weather condition.

7.2.4 Heavy construction for strong thermal inertia for walls and roofs: Time lag more than 8 hours

In the hot dry climate, high heat resistance and high heat capacity of the envelope elements are necessary. High resistance minimizes the conductive heat flow into the building mass during the daytime. Actually, this would reduce the rate of cooling the building mass during nighttime, but it could be overcome by employing

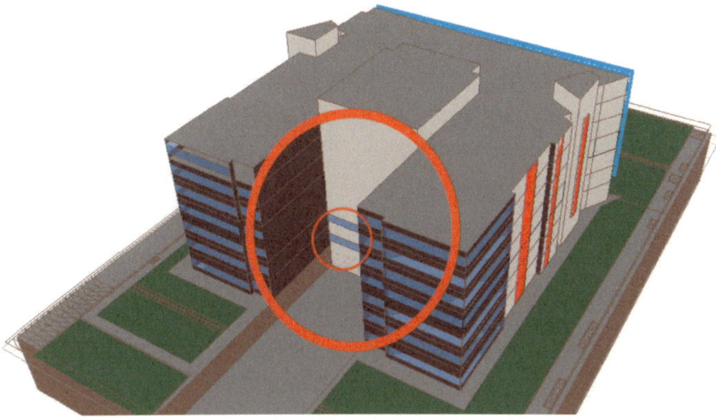

Figure 7.
Required opening between the courtyard and the backyard (done by the author after Autodesk Ecotect).

night purge ventilation strategy and new techniques of sunscreen which allow air movement [70, 75]. High thermal mass has been achieved traditionally by thick walls that are made of heavy materials such as stone, brick, adobe, and mud. To achieve this with the glazing wall, it has been modified to be double-tinted glazing. A U-value of 1.0 W/m^2 K has been used for the external facades. A canopy was added to the southern facade in the form of glazed sunscreen. Shading devices have been designed for the west facade to avoid the very hot solar rays of the afternoon. Firstly, a plan of blocking the solar rays of the summer season from 1:00 pm to 5:00 pm was achieved by 2.4 m depth shading device, which would not be accepted by the architectural consultant and the city council regulations. Therefore, the time range has been minimized to be between 1:00 pm and 3:00 pm and combined between the vertical and horizontal shading devices to minimize the depth of the devices to be 1.0 m (**Figure 8**). The same shading devices have been applied to the east facade for esthetic reasons.

7.2.5 Shaded roof

It also recommended to shade part of the roof, particularly the service area, with a pergola that can used for the photovoltaic cells to generate green power (**Figure 9**).

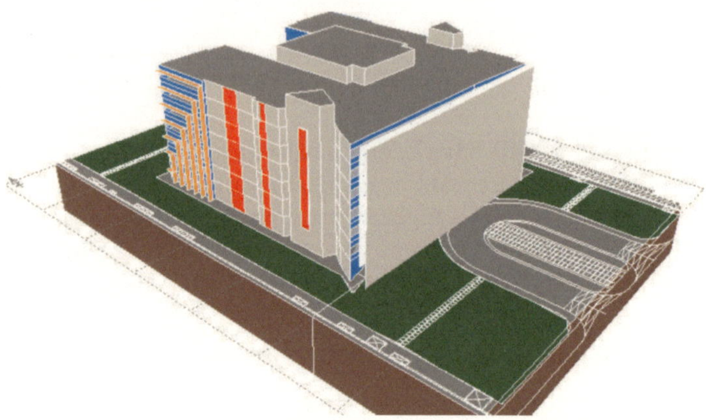

Figure 8.
Proposed vertical/horizontal shading devices on the west/east facades (done by the author after Autodesk Ecotect).

Figure 9.
Shading part of the roof (done by the author after Autodesk Ecotect).

7.2.6 Daylight and lighting views

To provide the building occupants with a connection to the outdoors, through the introduction of daylight and views into the regularly occupied areas of the building (**Figure 10**), it has been recommended to achieve a direct line of sight to the outdoor environment via vision glazing between 30 inches (75 cm) and 90 inches (225 cm) (**Figure 11**) above the finish floor for building occupants in 90% of all regularly occupied areas [73]. The floor area of the typical floor plan has been simulated using Autodesk Ecotect, and the daylight has been calculated on a height of 30 in (75 cm) above the floor. An interval of 5 foot (150 cm) has been employed in the analysis grid in the two directions X and Y. The first results did not satisfy the credit condition with the windows at sill height of 90 cm. Therefore, the height of the sill height was changed to be 30 in (75 cm).

By calculating the nodes of more than 25 fc (269.1 lux), the calculation showed that 472 out of 568 nodes are more than 25 fc and less than 500 fc. The percentage of area under the acceptable condition of the credit = 472/568 = 83.09% which is more than the required level by LEED (83.09% > 75%) (**Figure 12**).

7.3 Simulation results analysis

Using Autodesk Ecotect, the base case and the modified case have been modeled and simulated. The thermal performance of the third floor has been utilized for the

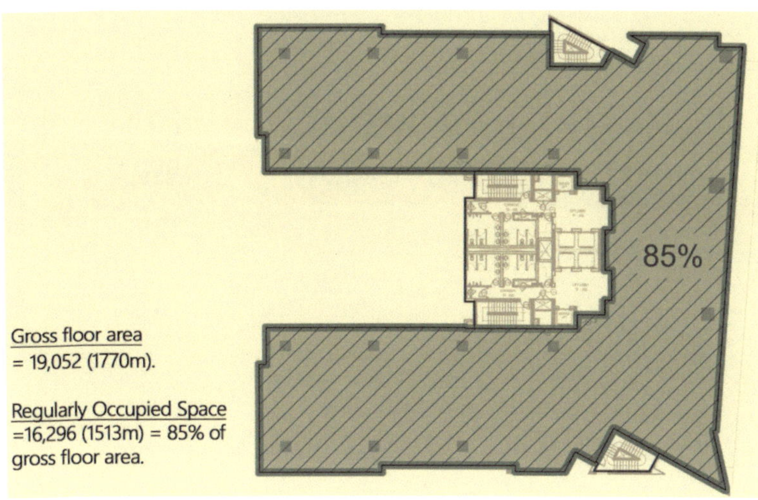

Figure 10.
Regularly occupied spaces to gross floor area (third floor) (done by the author).

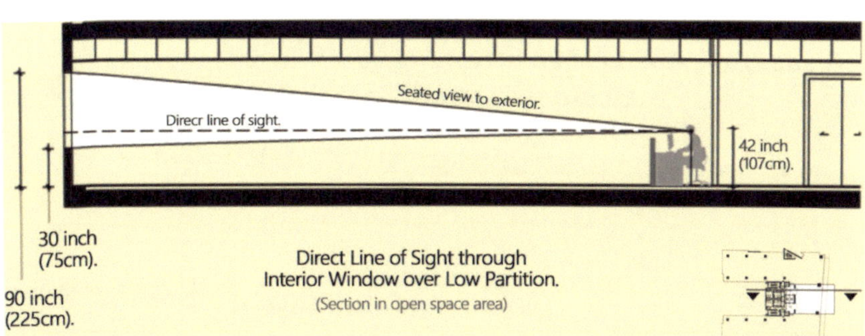

Figure 11.
Direct lines of sight to the exterior (done by the author).

comparison purpose. The same specifications of the zone in terms of air velocity, number of occupants, latent heat, operation hours, occupant activity and cloth, etc. were given for the two case scenarios. The passive heat gain breakdown of the building has been calculated for both the base case of the GREENEDGE building

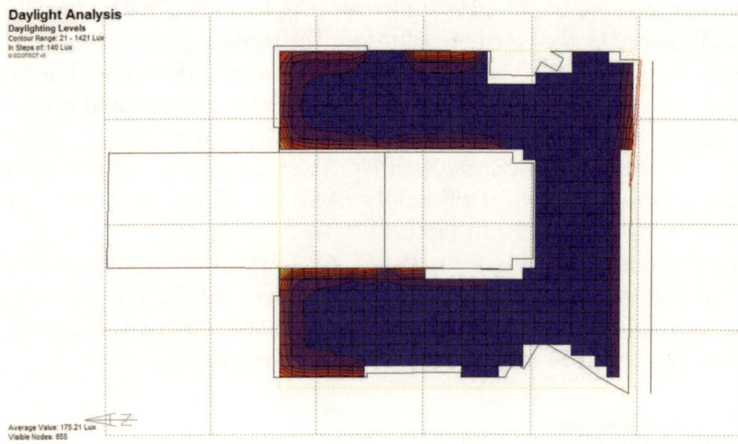

Figure 12.
Daylight levels at the third floor of the building on the 21st of September for all the occupied spaces (after Autodesk Ecotect).

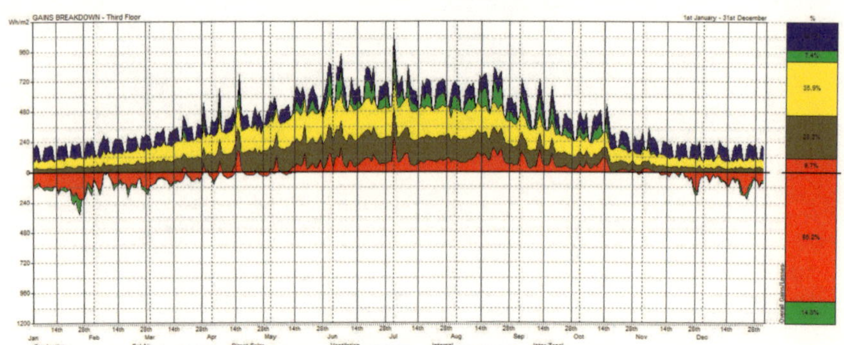

Figure 13.
Passive heat gain breakdown of the base case (done by the author using Autodesk Ecotect).

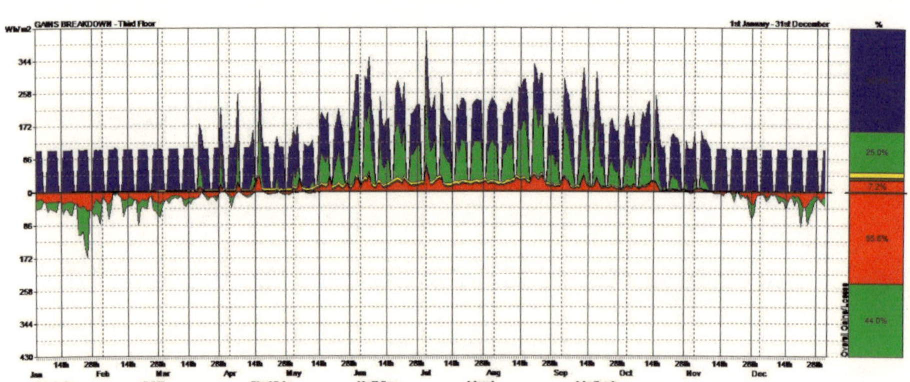

Figure 14.
Passive heat gain breakdown of the proposed case (done by the author using Autodesk Ecotect).

and the after modifications. **Figures 13** and **14** and show that the passive heat gain breakdown for the proposed case after modification was almost half the passive heat gain breakdown of the base case.

8. Conclusion

Although the GREENEDGE building is a mechanical-ventilated building (active ventilation), passive strategies and measures were followed to minimize the required energy for cooling and heating loads during the different seasons. The total hours of the years during energy consumption has been reduced by 12% after energy modeling analysis. The design recommendations could be summarized as follows:

1. Maximizing the amount of vegetation inside and outside buildings and using drip irrigation system to minimize the water consumption.

2. Having a northern courtyard with link between the courtyard and the external environment (the Takhtabush).

3. Shading part of the roof, particularly the service area, with a pergola that can be used for the photovoltaic cells to generate green power with the solar reflective index (SRI) not more than 29.

4. Placing vertical and horizontal shading devices on the west/east facades to block the solar radiation during the noontime of the day.

5. Windows has been modified to include two parts (lower and upper parts) that can give the occupants the controllability of opening the upper or the lower parts according to the weather condition. Those of the north direction must be the same in height with the human being. Opposite ones must be in a higher position to enable the required cross-ventilation.

6. Heavy construction for strong thermal inertia for walls and roofs: time lag more than 8 hours. To achieve this with the glazing wall, a doubled glazing with a U-value of 1.0 W/m^2 K has been used at the south and west facades.

7. A canopy has been added to the south facade in the form of Mashrabia, shading devices, or glazed screen.

The GREENEDGE building imitating the LEED goal for a golden certificate is packed with good design potentials which can lead for such project to be one of the first office buildings in Egypt to be certified with a Golden certification using the newly announced LEED for Core and Shell. It is worth mentioning here that the building has been achieved a Preliminary Platinum Certificate.

Acknowledgments

The author wishes to thank GREENEDGE owner, management bodies, and the design teams for their cooperation and efforts.

Author details

Mady Mohamed[1,2]

1 Effat University, Jeddah, KSA

2 Zagazig University, Zagazig, Egypt

*Address all correspondence to: momohamed@effatuniversity.edu.sa

References

[1] Nations U. World population projected to reach 9.7 billion by 2050. 2015. Available from: http://www.un.org/en/development/desa/news/population/2015-report.html

[2] UN-Habitat. Urban themes. 2012. Available from: https://unhabitat.org/urban-themes/climate-change/

[3] Hellström T. Dimensions of environmentally sustainable innovation: The structure of eco-innovation concepts. Sustainable Development. 2007;**15**(3):148-159

[4] Communities S. STAR + LEED FOR CITIES: STAR. 2019. Available from: http://www.starcommunities.org/

[5] Mahdavinejad M, Zia A, Larki AN, Ghanavati S, Elmi N. Dilemma of green and pseudo green architecture based on LEED norms in case of developing countries. International Journal of Sustainable Built Environment. 2014;**3**: 235-246

[6] Mohamed M, Gado T. Assessment of thermal comfort inside primary governmental classrooms in hot dry climates Part I: A case study from Egypt. In: Horner M, Price A, Bebbington J, Emmanuel R, editors. SUE-MoT 2009 Second International Conference on Whole Life Urban Sustainability and its Assessment. Loughborough, UK: Loughborough University; 2009. pp. 979-990

[7] Mohamed M, Gado T. Assessment of thermal comfort inside primary governmental classrooms in hot dry climates Part II: A case study from Egypt. In: Horner M, Price A, Bebbington J, Emmanuel R, editors. SUE-MoT 2009 Second International Conference on Whole Life Urban Sustainability and its Assessment. Loughborough, UK: Loughborough University; 2009. pp. 991-1001

[8] Vierra S. Green Buildings Standards and Certification Systems: National Institute of Building Sciences. 2016. Available from: https://www.wbdg.org/resources/green-building-standards-and-certification-systems

[9] Goals SD. Reference framework of sustainable cities (RFSC). 2008. Available from: https://urbact.eu/reference-framework-sustainable-cities

[10] UN-Habitat. City prosperity Initiative (CPI) Nairobi, Kenya. 2012. Available from: http://cpi.unhabitat.org/

[11] Group CCCL. Climate Positive Development Program. 2009. Available from: www.climatepositivedevelopment.org

[12] Programme TUNE. UNEP-SBCI—The United Nations Environment Programme - Sustainable Building and Climate Initiative. 2012. Available from: http://www.unep.org/sbci/index.asp

[13] GBA. Green Building Certifications, Rating Systems, & Labels Pittsburgh, PA. 2016. Available from: https://www.go-gba.org/resources/building-product-certifications/

[14] Bernardi E, Carlucci S, Cornaro C, Bohne R. An analysis of the Most adopted rating Systems for Assessing the environmental impact of buildings. Sustainability. 2017;**9**(7):1226

[15] Mohamed M, Al-kesmi S, AlSurf M, editors. Framework for successful development of green cities in Saudi Arabia. In: The 2nd International Memaryat Conference—Architecture and Urban Resiliency. Jeddah, KSA: Effat University; 2018

[16] Roderick Y, McEwan D, Wheatley C, Alonso C. A Comparative Study of Building Energy Performance Assessment between LEED, BREEAM

and Green Star Schemes Glasgow. Kelvin Campus: U.K Integrated Environmental Solutions Limited; 2010

[17] Gou Z, Lau SS-Y. Contextualizing green building rating systems: Case study of Hong Kong. Habitat International. 2014;**44**:282-289

[18] Asdrubali F, Baldinelli G, Bianchi F, Sambuco S. A comparison between environmental sustainability rating systems LEED and ITACA for residential buildings. Building and Environment. 2015;**86**: 98-108

[19] Waidyasekara KGAS, Silva MLD, Rameezdeen R. Comparative study of green Byilding rating systems: In terms of water efficiency and conservation. In: The Second World Construction Symposium: Socio-Economic Sustainability in Construction 14–15 June. Colombo, Sri Lanka; 2013. pp. 108-117

[20] Fowler KM, Rauch EM. Sustainable Building Rating Systems Summary. US: Pacific Northwest National Laboratory oftUSDoEbB; 2006

[21] Brodach M. Green Building Market Situation in Russia. 2011. Available from: http://zvt.abok.ru/articles/47/Green_Building_Market_Situation_in_Russia

[22] Ibrahim IAS. Green architecture challenges in the Middle East within different rating systems. Energy Procedia. 2017;**115**:344-352

[23] GBC F. InternatIonal envIronmental certIfIcatIons for the desIgn and constructIon of non-resIdentIal buIldIngs—The positioning of HQE certification realtive to BREEAM and LEED. 2015

[24] BRE. BREEAM UK: Building Research Establishment Ltd. 1990. Available from: https://www.breeam.com/

[25] Council USGB. Leadership in Energy and Environmental Design USA. 2000. Available from: https://new.usgbc.org/leed

[26] Counceil IGB. IGBC Rating Systems India. 2001. Available from: https://igbc.in/igbc/

[27] JSBC, Conservation (IBEC) IfBEaE. CASBEE (Comprehensive Assessment System for Built Environment Efficiency) Japan. 2001. Available from: http://www.ibec.or.jp/CASBEE/english/

[28] Japan Sustainable Building Consortium (JSBC), Conservation (IBEC) IfBEaE. Comprehensive Assessment System for Built Environment Efficiency. 2004. Available from: http://www.ibec.or.jp/CASBEE/english/

[29] Inc. GC. Green Globe USA: Green Globe Ltd. 2002. Available from: https://greenglobe.com/

[30] GBCA. The Green Star rating system Australia. 2002. Available from: https://new.gbca.org.au/green-star/rating-system/

[31] GBCI, Bob Galvin foM. Performance Excellence in Electricity Renewal (PEER) USA. 2003. Available from: http://www.gbci.org/press-kit-peer

[32] Authority BaC. BCA Green Mark Singapore. 2005. Available from: https://www.bca.gov.sg/greenmark/green_mark_buildings.html

[33] Communities S. Council USGB. STAR community rating system USA. 2007. Available from: http://www.starcommunities.org/

[34] GofRD. Global Sustainability Assessment System (GSAS): An Overview. 2014

[35] GOfRD. Global Sustainability Assessment System (GSAS). 2007.

Available from: https://www.gord.qa/gsas-trust

[36] Africa GBCS. Green Star Tools South Africa. 2007. Available from: https://gbcsa.org.za/certify/green-star-sa/

[37] Academy D. Council GSB. DGNB Global Benchmark for Sustainability Germany. 2007. Available from: https://www.dgnb-system.de/en/system/certification_system/index.php

[38] Larsson N, Part A. User Guide to the SBTool 2012 Assessment Framework. iiSBE; 2012

[39] (iiSBE) IIfaSBE. SBTooL - Sustainable Buildings Tool Lithuania. 2009. Available from: http://www.iisbe.org/sbmethod

[40] El-Demerdash M, Mosallam A, Bahnasawy H, Fakhry S, AlZahaby K. Road to Energy Efficient and Environmentally Friendly Affordable Construction Systems. Egyptian Green Building Council; 2009

[41] EGBC. Green Pyramid Rating System (GPRS) Egypt. 2009. Available from: http://egypt-gbc.org/ratings.html

[42] Council NZGB. Green Star—NZ New Zealand. 2009. Available from: https://www.nzgbc.org.nz/GreenStar

[43] Limited HKGBC. HK-BEAM - Building Environmental Assessment Method Hong Kong. 2009. Available from: http://greenbuilding.hkgbc.org.hk/

[44] Lanka GBCoS. GREENSL® Rating System for Built Environment Sri Lancka. 2010. Available from: http://srilankagbc.org/rating.php#

[45] Bhd GS. Green Building Index (GBI) Malaysia. 2010. Available from: http://new.greenbuildingindex.org/

[46] Council G. Green Rating for Integrated Habitat Assessment (GRIHA) India. 2010. Available from: http://www.grihaindia.org/griha-rating

[47] Council ADUP. Pearl—BRS—Pearl Building Rating System (PBRS) Abu-Dhabi. 2010. Available from: http://www3.cec.org/islandora-gb/en/islandora/object/greenbuilding%3A101

[48] Council SGB. Miljöbyggnad Sweden. 2012. Available from: https://www.sgbc.se/certifiering/miljobyggnad/

[49] Efficiencies E-EiDfG. EDGE User Guide, Version 2.1 In: Corporation IF, editor. 2018

[50] Program IsGBMT. Excellence in Design for Greater Efficiencies (EDGE): EDGE, International Finance Corporation, World Bank Group. 2012. Available from: https://www.edgebuildings.com/marketing/edge/

[51] LGBC. ARZ Green Building Rating System Lebanon. 2012. Available from: http://arzrating.com/

[52] HAAplHQEA. HQE_France's reference certification scheme France. 2013. Available from: https://www.behqe.com/

[53] IWBI. The WELL Building Standard® USA: U.S. Green Building Council. 2015. Available from: https://www.usgbc.org/articles/what-well

[54] BRE. An introduction to CEEQUAL—Improving sustainability through best practice. UK: BRE Trust. 2018

[55] BREL. CEEQUAL UK. 2015. Available from: http://www.ceequal.com/

[56] Mohamed M. An approach to integrate the environmental impact assessment process in the early stages of design. In: The First International Engineering Conference Hosting Major International Events Innovation,

Creativity and Impact Assessment 15–18-January Housing & Building National Research Center. Cairo, Egypt: HBRC; 2013

[57] Mohamed M, Gado T. Application of computer based environmental assessment and optimization tools: An approach for appropriating buildings. In: 3rd International Conference ArchCairo 2006, Appropriating Architecture Taming Urbanism in the Decades of Transformation; 21–23 Feburary 2006. Cairo; 2006. pp. 592-604

[58] Mohamed M. Investigating the environmental performance of Government primary schools in Egypt: with particular concern to thermal comfort [PhD thesis]. Dundee: Dundee University; 2009

[59] American Society of Heating RaA-CE, Inc. Advanced energy design guide for K-12 school buildings. 2008

[60] U.S. Green Building Council. LEED for Schools U.S. Green Building Council (USGBC). 2008. Available from: http://www.usgbc.org/DisplayPage.aspx?CMSPageID=1586

[61] U.S. Green Building Council. LEED for new construction and major renovations. U.S. Green Building Council (USGBC). 2009

[62] Mohamed M. Traditional ways of dealing with climate in Egypt. In: Lehmann S, Waer HA, Al-Qawasmi J, editors. The Seventh International Conference of Sustainable Architecture and Urban Development (SAUD 2010); 12–14, July. Amman, Jordan: The Center for the Study of Architecture in Arab Region (CSAAR Press); 2010. pp. 247-266

[63] Mohamed M, Anwar M, Okasha S. Computer aided design software in site planning. In: Al-Azhar Engineering Eleventh International Conference; 21—23 Dec-2010. Cairo, Egypt; 2010

[64] Mohamed M, Gado T. Investigating the process of exporting autodesk ecotect models to detailed thermal simulation software. Environment and Ecology Research. 2014;**2**(5):209-220

[65] Mohamed M, Gado T, Unwin S. The environmental performance of classrooms in Egypt: A case study from El-Minya governorate. In: Egbu PCO, Tong MKL, editors. The Second Scottish Conference for Postgraduate Researchers of the Built & Natural Environment (PRoBE 2005); 16–17 November 2005. Glasgow Glasgow Caledonian University; 2005. pp. 643-651

[66] Mohamed M, Gado T, Osman M. Investigating the intelligence of the low-tech earth architecture of the Sahara: A feasibility study from the western desert of Egypt. Intelligent Buildings International. 2010;**2**(2):179-197

[67] Mohamed M. Lessons from the past to enhance the environmental performance of primary school classrooms in Egypt. Environment and Ecology Research. 2014;**2**(6):221-233

[68] Maps G, cartographer New Cairo, Egypt. 2015

[69] Councile NCC. New Cairo Map. Cairo. 2013

[70] Givoni B. Climate Consideration in Building and Urban Design. New York, USA: Van Nostrand Reinhold; 1998

[71] Dixon AM, Butler D, Fewkes A. Guidelines for grey water re-use: Health issues. Water Environment Journal. 1999;**13**(5):322-326

[72] Hespanhol I. 1 AMEP. Who guidelines and national standards for reuse and water quality. Water Research. 1994;**28**(1):119-124

[73] U.S. Green Building Council. Green Building Design and Construction.

LEED Reference Guide for Green Building Design and Construction. Washington, DC: U.S. Green Building Council (USGBC); 2010. p. 2010

[74] Fathy H. Natural Energy and Vernacular Architecture: Principles and Examples with Reference to Hot Arid Climates. University of Chicago Press; 1986

[75] Givoni B. In: Cowan HJ, editor. Man, Climate and Architecture. 2nd ed. London: Applied science publishers LTD; 1976. p. 483

Chapter 8

The Synergistic Impact of Climate Change and Anthropogenic Management on the Lake Kinneret Ecosystem

Moshe Gophen

Abstract

Anthropogenic and natural processes caused significant changes in Lake Kinneret and its drainage basin ecosystems. Climate change of warming and dryness induced a decline in the lake water level. Changes in the composition structure of the phytoplankton assemblages were enhanced by the decline of nitrogen availability resulting in reduction of Peridinium and enhancement of Cyanobacteria. Increase of the phosphorus availability enhanced Chlorophyta and diatoms. Nutrient export from the Hula Valley to Lake Kinneret is discharge-dependent. The external input decline of organic nitrogen and total dissolved phosphorus is due to anthropogenic achievements. Nitrogen decline and slight increase of phosphorus in Lake Kinneret were followed by Peridinium decline and increase of non-Peridinium algae. The resulting change of food-web structure and water quality in the Kinneret was a shift from phosphorus to nitrogen limitation, which enhanced cyanobacteria.

Keywords: Kinneret, Hula Valley, Peridinium, cyanobacteria, climate change

1. Introduction

1.1 The Kinneret drainage basin

The Kinneret drainage basin (total area 2730 km^2) is located mostly northern to the lake (**Figure 6**). Its maximum length from north to south is 110 km and widest 50 km. The Kinneret drainage basin is bounded between the Mount Hermon block and Litani basin in the north, Meron and Naftali mountain ridges in the west, the Golan Heights in the east, and Kinarot Valley in the south. The highest altitude point within the Kinneret basin (Hermon summit) is 2814 m (ASL). Taking into account Kinneret common water level (WL) as 213 m (BSL), the total slope of 3027 m along 110 km makes the mean surface gradient as 2.8%. The Hula Valley is flat with a mean slope of appox. 0.7%. The Kinneret basin comprises the following subunits: northern basin including Jordan drainage basin drained by the rivers: Hermon, Dan, and Snir, and the Hula Valley-1530 km^2 (56%); eastern basin drained by several rivers (Meshushim, Daliyot, Zavitan, and others)-580 km^2 (21%); western basin drained by the rivers of Zalmon and Amud—450 km^2 (16%); and several other small parts southern and western to the lake.

1.2 A brief historical survey of the Hula Valley management

During the last 80 years, the Lake Kinneret drainage basin ecosystems have undergone significant anthropogenic and natural modifications. Prior to the 1950s, the Hula Valley was mostly (6500 ha) covered with old Lake Hula (1300 ha) and swampy wetlands. This area was not cultivated, malaria was common, and water loss by evapotranspiration (ET) was significant. The Jordan River crossing the Hula Valley contributes about 63% of the downstream of the Lake Kinneret's water budget, but 70% of the total nutrient inputs, of which over 50% originate in the Hula Valley region, including the valley and the slopes on both sides (east and west) of it. Old Lake Hula and swamps were drained and were being converted for agricultural development. Years later, land utilization was modified in an operation referred to as the Hula Reclamation Project (HRP) which was improved later. The HRP included creation of a new shallow lake Agmon (surface area of 1120 ha, mean depth—0.45 m., volume—0.44×10^6 m^3), renewal of 90 km drainage and water supply canals, placing a vertical plastic barrier along 2.8 km crossing the valley from east to west, maintenance of higher underground water table, and functional conversion of 500 ha with lake Agmon in the center from agricultural to eco-tourism usage. The objectives of HPR were aimed at: (1) nutrient removal from the Lake Kinneret external loads through the Lake Agmon hydrological system; (2) to produce an ecological component for eco-tourism-Lake Agmon; and (3) the usage of Lake Agmon as a principal component for the hydrological management and agricultural irrigation system for the entire valley. The following objectives were implemented: improvement of irrigation water supply, maintenance of high underground water table ensuring peat soil moisture to prevent its deterioration, and the achievement of a high diversity of re-establishment of natural flora and fauna emphasizing aquatic birds.

Prior to the drainage of old Lake Hula and adjacent swamps during 1950–1957, Nitrogen was fluxed from the basin to the lake, mostly as highly bioavailable ammonia, but after the Hula drainage, the dominant N was modified to nitrate. Before the mid-1990s, a daily volume of 25×10^3 m^3 of raw sewage and fishpond (1700 ha) effluents rich with ammonia fluxed into Lake Kinneret. The fishpond area was dramatically reduced (450 ha), as well as its effluents, and the raw sewage was stored in reservoirs and reused. As a result of inappropriate irrigation and agricultural methods, the peat soil quality deteriorated by consolidation, destruction, and surface subsidence. It was accompanied by heavy dust storms, blocking of drainage canals, enhancement of underground fires, and outbreaks of rodent populations. These deteriorated processes caused severe damage to agricultural crops. A reclamation project (HRP) was consequently implemented.

The aim of the present paper is to evaluate the outcome of the anthropogenic intervention in the Hula Valley, together with climate change, on the management of Lake Kinneret, its water quality, and the entire ecosystem structure trait.

2. Methods

Climatological (precipitation, air temperature) data was supported by the Israeli Meteoroloical Service. Jordan River discharge and nutrient concentrations were supported by Kinneret Limnological Labortory Data Base and Mekorot Water Supply Company Jordan District. The nutrient concentrations and the phytoplankton composition structure in the epilimnion of Lake Kinneret during 1970–2018, as well as water level and ground water table (GWT) in the Hula Valley were all reevaluated and statistically analyzed. The information was given by Kinneret

Limnological Laboratory (IOLR), Mekorot-Water Supply Company-The Jordan Region Monitor Unit, The Israeli Hydrological Service, National Water Authority, and the Israeli Meteorological Service. The Statistical analyses used were fractional polynomial and linear regressions. The data on Agmon WL and the GWT depths were supplied by the Data Base Center of the Hula Project in the MIGAL-Scientific Research Institute.

3. Results

3.1 Climate change

The periodical occurrence of drought is given in **Table 1**. Results in **Table 1** indicate 62% periods (years) of three levels (A, B, and C) drought as SPI (specific precipitation index) values [1] during 1930–1980, and 42% during 1981–2014. Moreover, percentage of normal conditioned periods was higher during the latter period. Nevertheless, recent 5 years (2014–2018) were consecutive drought seasons.

Data shown in **Figure 3** indicate precipitation decline since mid-1980s. The information that is evaluated in **Figure 1** was collected (from 1940) in the Dafna Station located in northern Hula Valley statistically indicating a decline of appox. 140 mm annually.

Periodical means of air temperatures measured in the meteorological Station in Dafna (Northern Hula Valley) during 62 years (1946–2008) resulted (°C) in the following changes: 20.2, 19.4, 18.9, and 19.8 in 1946–1958, 1959–1982, 1983–1990, and 1991–2008, respectively, indicating fluctuation ranges of −0.8, −0.5, and +0.9°C.

SPI level	1930–1980	1981–2014
A	34	21
B	26	18
C	2	3
D	38	58

A—close to normal conditions; B—moderate drought; C—severe drought; and D—normal conditions.

Table 1.
Drought level (A, B, C, and D) occurrence (%) evaluated as SPI values [1].

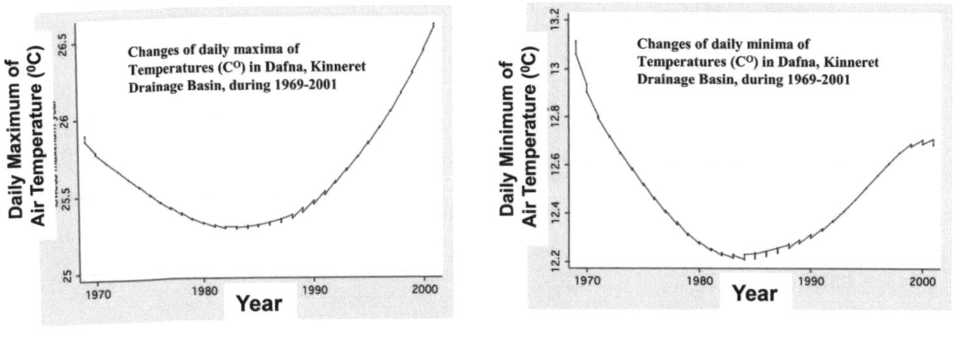

Figure 1.
Trend of changes (LOWESS; 0.8) of daily maxima (left panel) and minima (right panel) of air temperature (°C) Dafna Station (Northern Hula Valley) during 1969–2001.

LOWESS smoothing statistical method provide locally weighted scatterplot smoothing. The smoothed values are obtained by running a regression of Y and X variables weighted where the central value gets the highest weight and other points around receive less weight. Moreover, in relation to those documented air temperature change, the lake water temperatures were fluctuated as well (**Figure 2**): The upper epilimnetic layer (0–10 m) became warmer since the early 1980s until the early 2000s by 1.9 (21.7–23.6), whilet the temperature increased in the lower layer (32 m deep) similarly by app 2.0°C but reasonably later (from late 1980s) (**Figure 2**).

A significant evidence for climate change is given in **Figure 3**, which represents a mean increase of 60 mm from 1940 until the mid-1980s and later a decline of app. 130 mm until present.

3.2 Hula reclamation project

The history and implementation of the Hula Project are given in [2–4]. Long-term data about the underground water table (GWT) in the Hula Valley are presented in **Figures 4** and **5**. The impact of drought condition is reflected from the

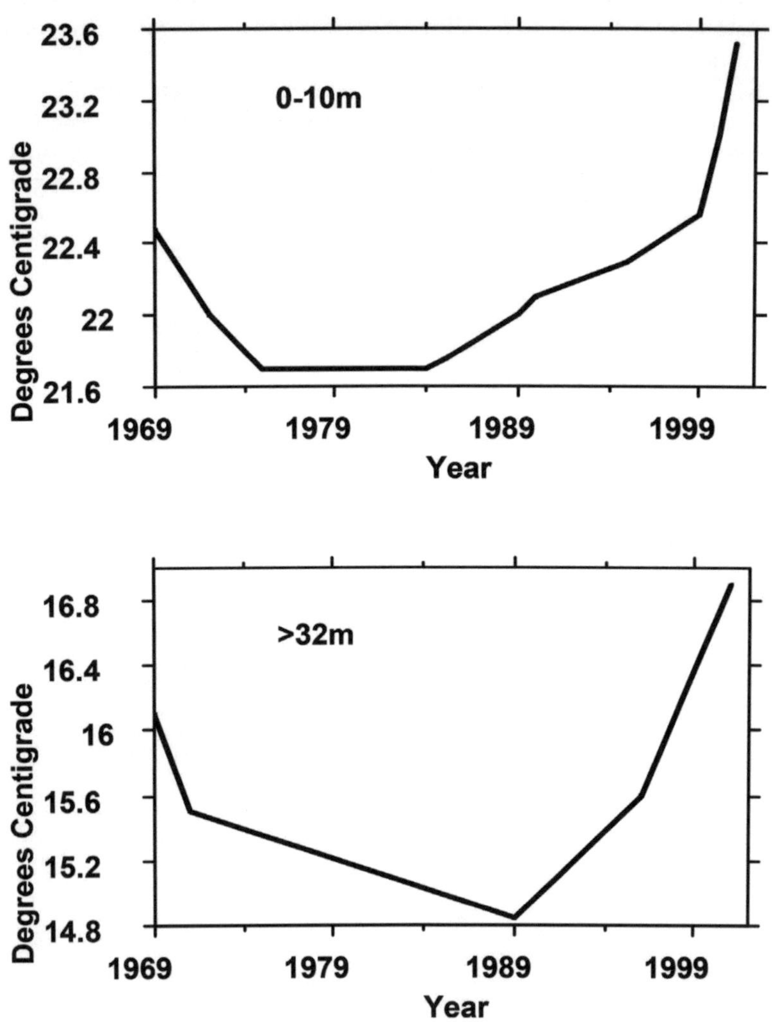

Figure 2.
Annual averages of water layers (0–10 m upper, >32 m lower) in Lake Kinneret during 1969–2001.

deepening of averaged GWT by 30 cm during recent 6–7 years (**Figure 4**). The data presented in **Figure 5** emphasize the underground topographical structure and the impact of the recent drought: the GWT altitude in the northern region of the Hula Valley is higher than that of the southern part. Therefore, hydrological gradient existent induces underground water migration from north to south. Hydraulic forces from north to south induce water flow aimed at south but no such gradient on east-west directions (**Figure 5**).

Significant relations between nutrient inputs from the upper Jordan River watershed into Lake Kinneret and the river discharge were indicated, and the qualitative results are given in **Table 2**.

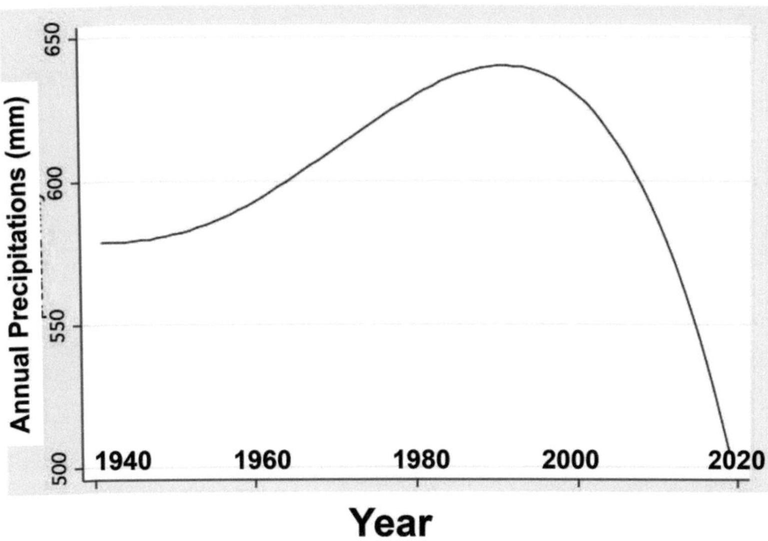

Figure 3.
Fractional polynomial regression between annual (1940–2018) precipitation and years.

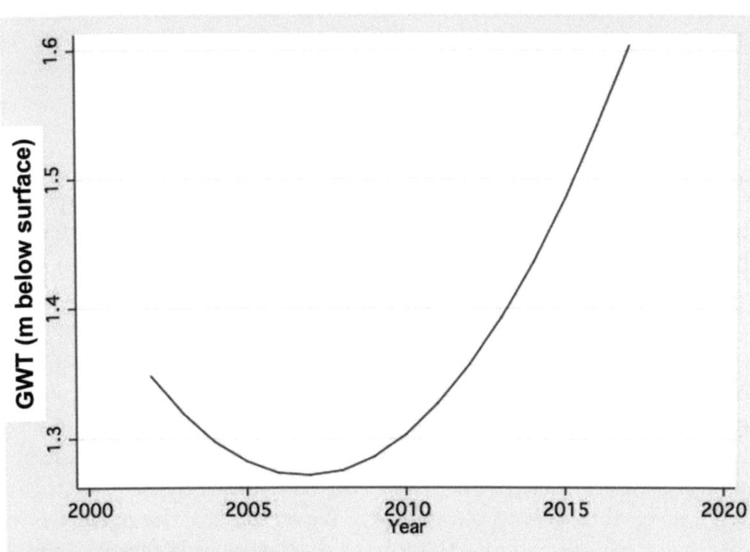

Figure 4.
Annual total (three Hula Valley regions) average of GWT (m below surface) during 2002–2018.

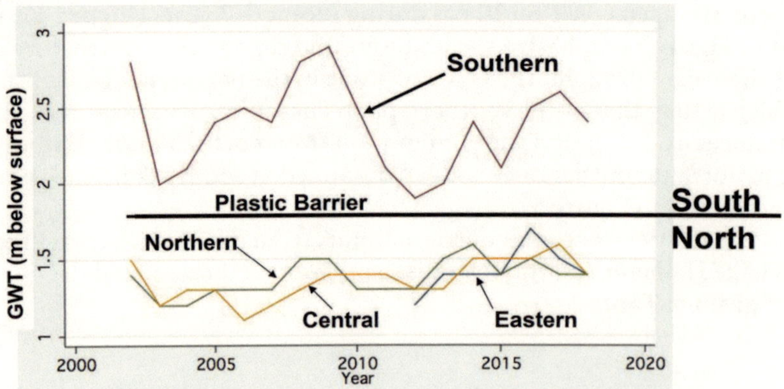

Figure 5.
Line scatter plot of annual (2002–2018) means of GWT (m below surface) in four Hula Valley regions: northern, eastern, western, and southern.

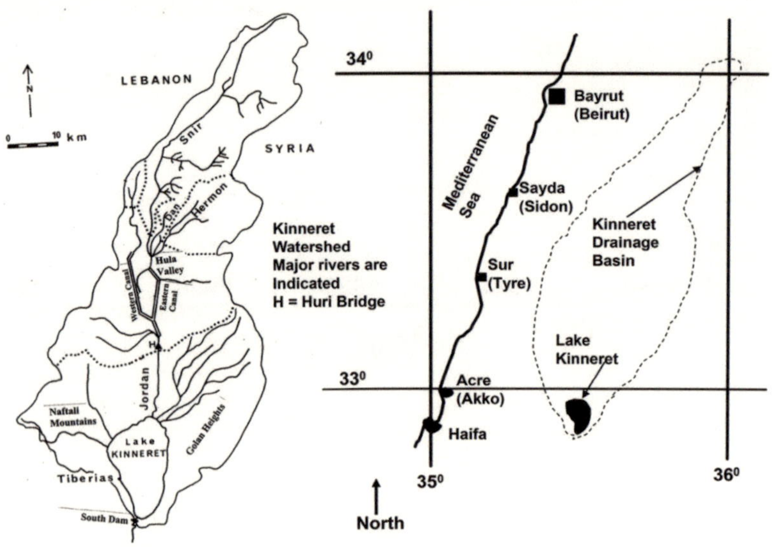

Figure 6.
Regional chart of Lake Kinneret watershed.

The quantitative significant relation between Jordan River Discharge and nutrient loads is obvious and was earlier documented, while data given in **Table 2** indicate the positive significant relation between the river discharge and the nutrient concentrations: the higher the discharge, the higher the nutrient concentration and quantities. The Linear regressions between Jordan discharge and nutrient loads is positively significant as presented in **Table 3**.

Results in **Tables 2** and **3** are compatible and strongly support the statement about positive linear regression between Jordan discharge and nutrient load transport into Lake Kinneret. These linear relations and the temporal decline of Jordan discharge are presented in **Figures 7–12** for organic nitrogen, total nitrogen, total phosphorus, and total dissolved phosphorus. Nevertheless, the pattern of relation between nitrate concentration and the Jordan discharge is different (**Figure 11**): NO_3 concentration increases in relation to time (from 1970 to 2018) and decreases relative to Jordan decline below 300 mcm/y and increases very little when discharge is elevated above 300 mcm/y. The different behavior of nitrate was studied by Geifman (1981) and the results are given in **Figure 12**. The significant increase of

Nutrient	r^2	p
Total nitrogen	0.1383	0.0046 S
Total phosphorus	0.4599	<0.0001 S
Nitrate	0.2012	0.0029 S
Ammonium	0.2527	0.0007 S
Organic nitrogen	0.5984	<0.0001 S
Total dissolved phosphorus	0.2019	0.0028 S
Kjeldahl total	0.6586	<0.0001 S
Kjeldahl dissolved	0.6417	<0.0001 S
Jordan discharge (mcm/y)	0.2004	0.0030 S

Table 2.
Results of linear regression analysis between Jordan River discharge and the multiannual means of nutrient concentrations (ppm): r^2, probability (p) values, and significance (S = significant) values are indicated. The annual Jordan River discharge (mcm/y; 10^6 m^3/y) is given.

Nutrient	r^2	p
Total nitrogen	0.860	<0.0001
Total phosphorus	0.596	<0.0001
TIN (total inorganic nitrogen)	0.776	<0.0001
Sulfate	0.816	<0.0001
Organic nitrogen	0.606	<0.0001
Chloride	0.886	<0.0001

Table 3.
Linear regressions between Jordan River discharge and nutrient loads (tons) r^2 are given and all probabilities were significant (<0.0001).

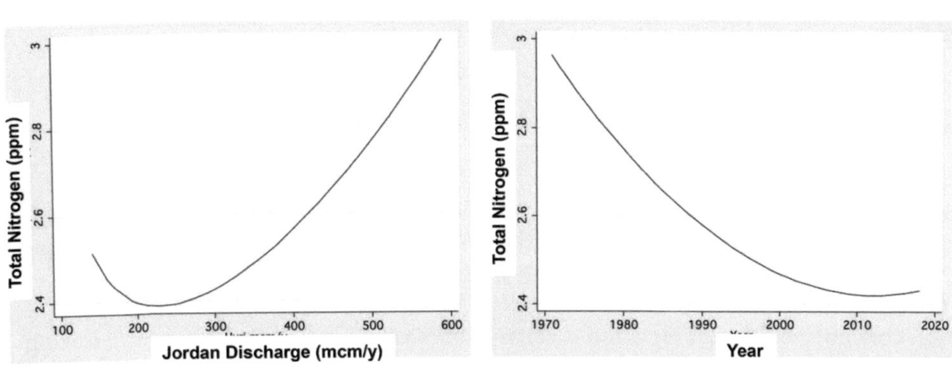

Figure 7.
Fractional polynomial regression between annual mean of the total nitrogen concentration (ppm) in Jordan water and Jordan water yield (mcm/y) (left) and with years (1970–2018) (right).

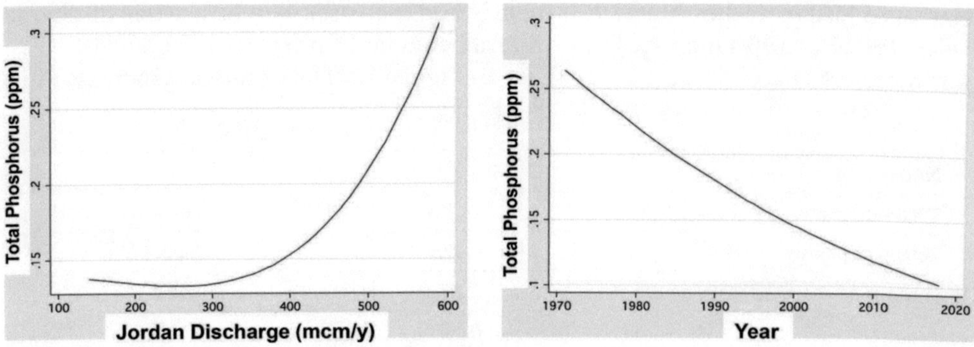

Figure 8.
Fractional polynomial regression between annual mean of the total phosphorus concentration (ppm) in Jordan water and Jordan water yield (mcm/y) (left) and with years (1970–2018) (right).

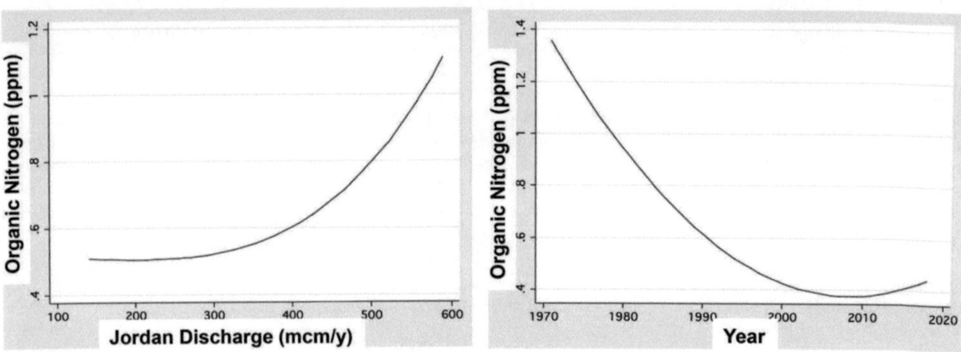

Figure 9.
Fractional polynomial regression between annual mean of the organic nitrogen concentration (ppm) in Jordan water and Jordan water yield (mcm/y) (left) and with years (1970–2018) (right).

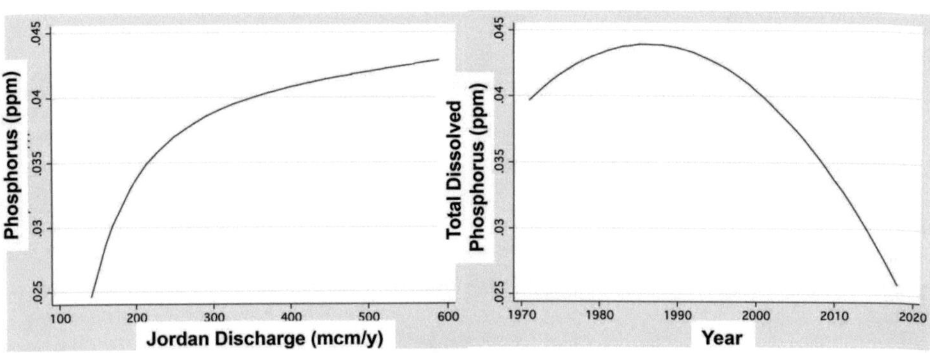

Figure 10.
Fractional polynomial regression between annual mean of the total dissolved phosphorus concentration (ppm) in Jordan water (left) and Jordan water yield (mcm/y) (left) and with years (1970–2018) (right).

NO_3 concentration with elevated discharge above 300 mcm/y as monitored during 1970–2018 is indicated (**Figure 11**). Nevertheless, the dynamic of nitrate flux from the peat soil in the Hula Valley is highly precipitation-dependent. The oxygenation process of the nitrogen rich peat soil produces latent nitrate which is easily released and migrates by precipitation water fluxes. Therefore, the relation between winter rain and nitrate concentration is highly positive. A study carried out in the winters

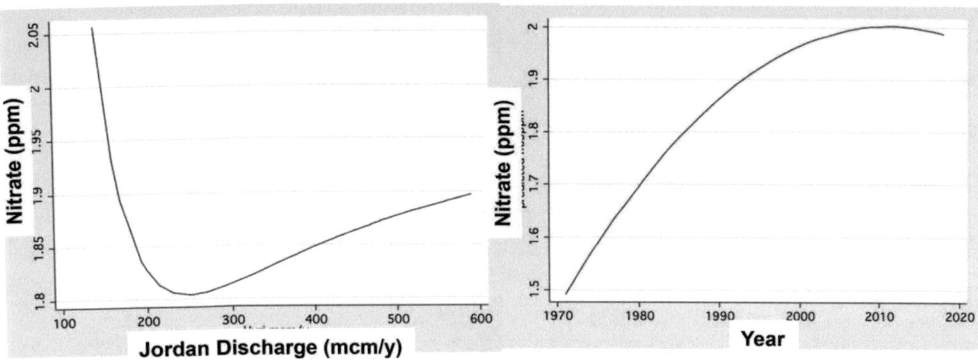

Figure 11.
Fractional polynomial regression between annual mean of the Nitrate concentration (ppm) in Jordan water and Jordan water yield (mcm/y) (left) and with years (1970–2018) (right).

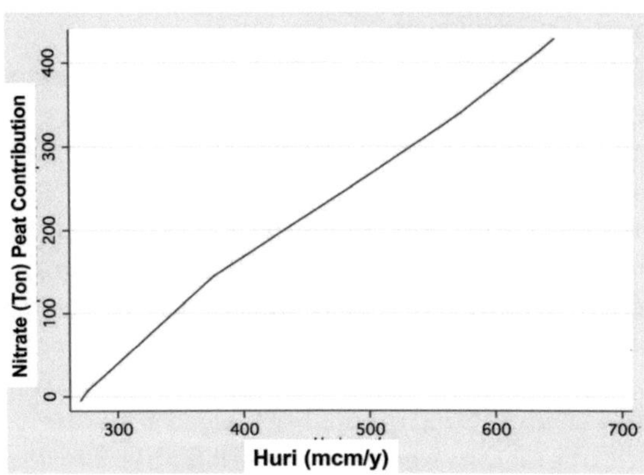

Figure 12.
Linear regression between annual loads (tons) of nitrates contributed by the organic-peat soil of the Hula Valley and the river Jordan annual water yields (m^3/mcm/y; 106 year) during 1969/70–1980/8 (Geifman, 1981, unpublished data).

	Month					
	January		February		March	
	Station					
Year	**Huri**	**Josef**	**Huri**	**Josef**	**Huri**	**Josef**
1975	1.30	1.28	1.77	1.49	1.56	1.23
1976	1.47	1.40	1.38	1.30	1.45	1.18
1977	1.90	1.41	1.88	1.18	1.43	1.18
1978	2.15	1.28	1.74	1.16	1.43	0.99
1979	1.45	1.40	1.57	1.23	1.43	1.19
1980	2.61	1.58	2.25	1.18	3.68	1.21
1981	4.71	1.57	4.49	1.35	2.20	1.11
Average	2.23	1.42	2.15	1.27	1.88	1.56

So far, any concentration elevation in the southern station is due to peat soil contribution of nitrate.

Table 4.
Monthly (rainy season: January–March) mean concentrations (ppm) in two sampling stations on the Jordan River: "Josef Bridge" located before crossing the Hula Valley and "Huri Bridge" at the southern part of the region after leaving the Hula valley.

Nutrient	Inlet		Outlet		Balance (tons/y)
	ppm	tons/y	ppm	tons/y	
TP	0.10	0.8	0.20	1.2	−0.4
TDP	0.04	0.2	0.04	0.2	0
TN	17.8	38.1	6.5	27.3	+10.8
TDN	17.4	34.7	6.4	21.6	+13.1
NO_3	11.8	19.4	5.3	3.3	+16.1
NH_4	5.3	10.1	1.2	7.2	+2.9

Table 5.
Annual mass balance (input minus output) (tons/y) of nutrients in Lake Agmon system during 2005.

during 1875–1981 (Geifman, 1981) documented the relative contribution of nitrate by the Hula Valley Peat soils (**Table 4**).

Results shown in **Table 4** indicate a supplemental mean peat soil contribution of nitrogen concentration (ppm) of 0.32 (March), 0.88 (February), and 0.83 (January). The Nitrate increment decline from January high precipitation gauge to lower rain regime in March is prominent.

The major objectives of the Hula Reclamation Project (HRP) were to ensure agricultural beneficiary and protect the quality of Lake Kinneret water. The achievement of agricultural development improvement was successfully accomplished; nevertheless, the removal of polluted nutrients was summarized as insignificant. Nitrogen removal was achieved mostly through the restriction of fishponds and sewage treatment; both were separately operated but not through the HRP operation. Results given in **Table 5** indicate the minor impact of Lake Agmon System (HPR) on polluted nutrient removal from the Lake Kinneret input loads. Kinneret Nitrogen sources are mostly external from the watershed and after restriction of fishpond and domestic sewage removal significant loads of organic Nitrogen were removed. Phosphorus supplemental resources to Lake Kinneret are dust deposition and lake bottom release; therefore, sewage removal and fishpond restriction did not lower supplement flux into lake water.

Results in **Table 5** indicate Lake Agmon system functioning as a sink for retained Phosphorus and removal of minor loads of nitrogen. The source of supplemental TP concentration in the Lake Agmon ecosystem is probably submerged plant mediated P intake from bottom sediments. The removal of 10.8 tons of TN is probably due to de-nitrification and sedimentation processes. Earlier studies documented the following long-term changes in the epilimnion of Lake Kinneret.

The cyanobacteria: Chlorophyta and diatoms proliferated and the dominance was shifted from large cell bloom forming dinoflagellate Peridinium gatunenze to cyanobacteria phytoplankton dominance where some of them are nitrogen fixers and the majority non-Nitrogen fixers. The nutritional structure of the Kinneret Epilimnion shifted from phosphorus to nitrogen limitation.

3.3 The impact of climate change on Lake Kinneret ecosystem

The decline of water level (WL) in Lake Kinneret (**Figures 13** and **14**) is an obvious result of climate change (precipitation and discharge decline). Nevertheless, it is likely that agricultural consumption might be an impact factor as well. Therefore, this anthropogenic factor was studied. The following data confirmed that anthropogenic usage of headwater discharges in the upper Jordan watershed until 1985 water consumption was declined from 100 to 120 mcm/y to 85 in 2015 and continuation of restriction came down to 68 mcm/y in 2018. The only significant long-term change of agricultural water usage in the Upper Jordan watershed is significant reduction. The outcome of WL decline was decline of nutrient content in the epilimnion of Lake Kinneret (**Figures 15** and **16**): the lower the WL, the lower the nutrient capacities in the epilimnion. Nevertheless, water quality implication is attributed to phytoplankton assemblages and the interrelations between

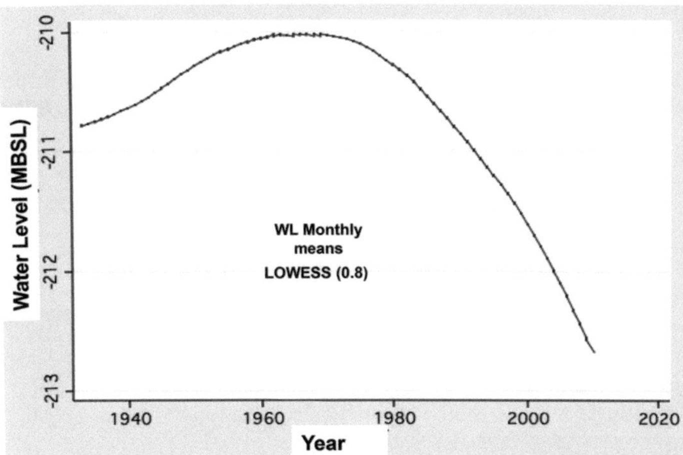

Figure 13.
Trend of changes (LOWESS 0.8) of monthly means of water level (MBSL) in Lake Kinneret during 1934–2018.

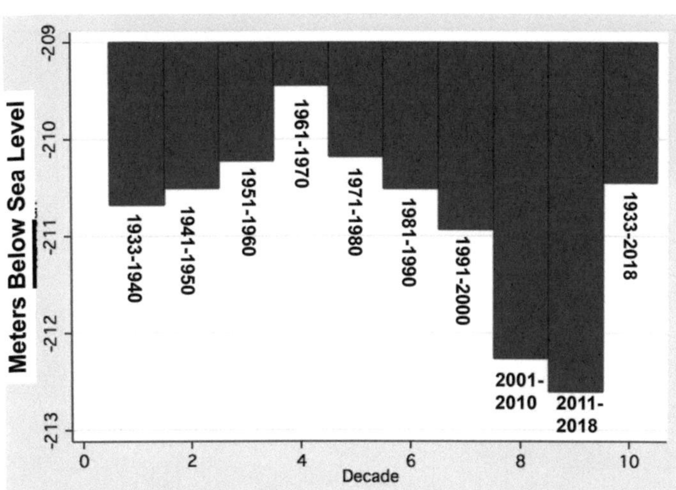

Figure 14.
Decade means of monthly averages of daily measured water level (MBSL) in Lake Kinneret during 1933–2018; decade no. 10 is total average.

nitrogen and phosphorus. **Figure 17** represents significant decline of nitrogen and a slight increase of the epilimnetic loads. The outcome was decline of the mass ratio between N and P from 70 to 23 (**Figures 17** and **18**). Such conditions are known as favored by nitrogen toxic cyanobacteria (17). Moreover, the insufficiency of

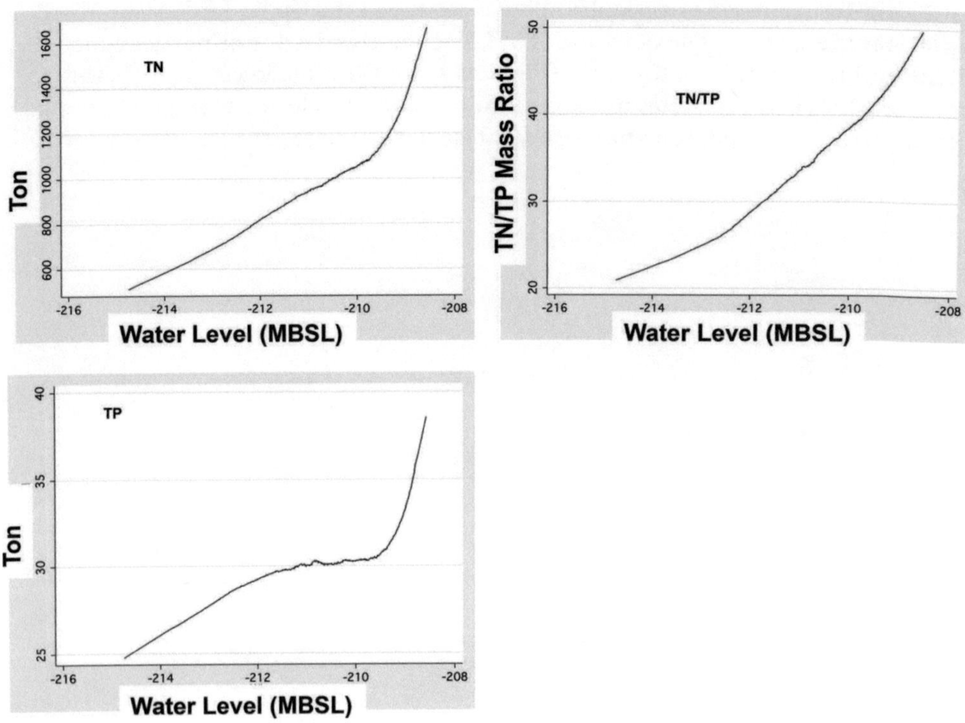

Figure 15.
Trend of changes (LOWESS 0.8) of epilimnetic load (T) of TN (upper left) and TN (lower left) and the TN/ TP mass ratio (upper right in relation to water level (MBSL)) decline during 1969–2001.

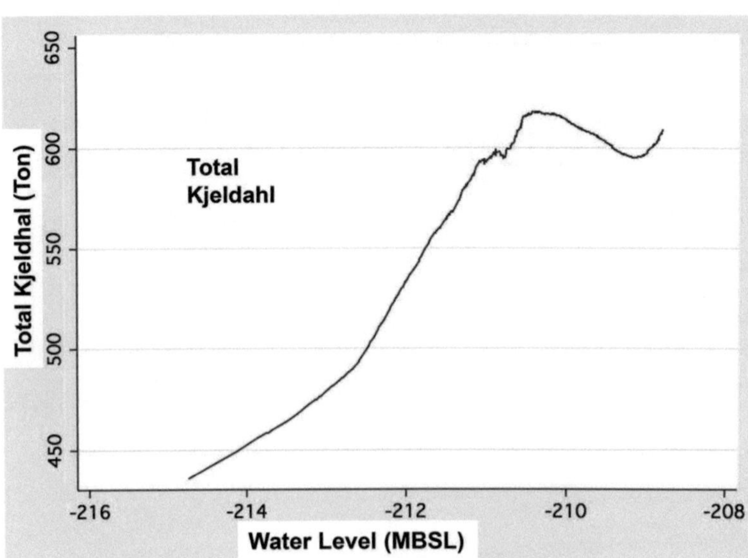

Figure 16.
Trend of changes (LOWESS 0.8) of epilimnetic load (T) of total Kjeldahl in relation to water level (MBSL) decline during 1969–2001.

nitrogen is unfavored by the long-term dominant bloom forming dinoflagellated Peridinium gatunenze and cyanobacteria became dominant accompanied by enhancement of Chlorophyta and diatoms (**Figures 19** and **20**).

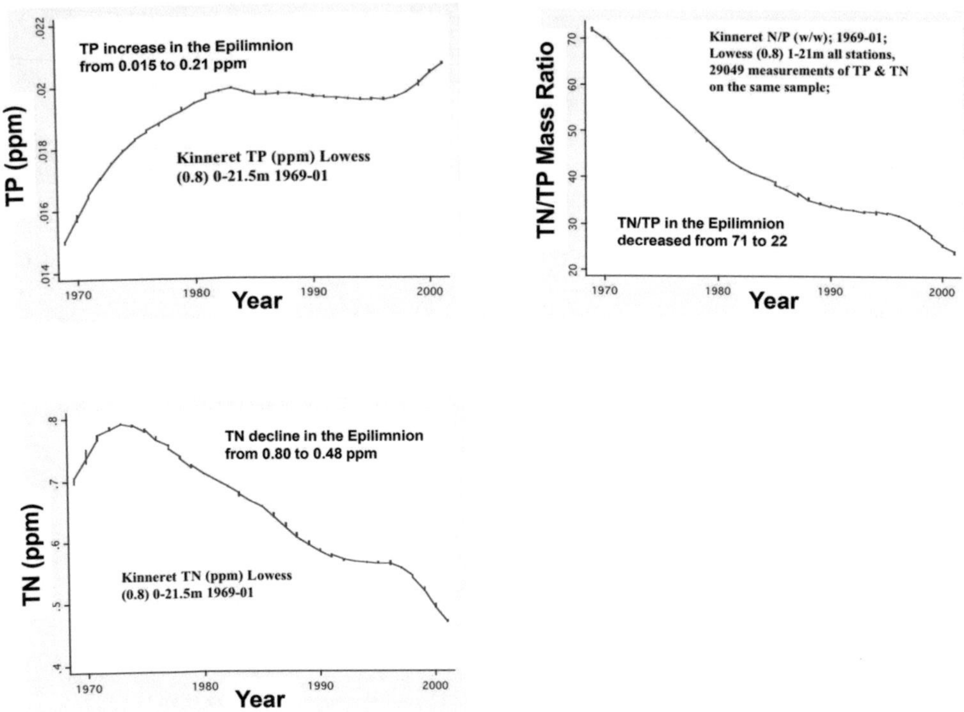

Figure 17.
Trend of temporal changes (LOWESS 0.8) of the concentration (ppm) of TP (upper left), TN (lower left), and TN/TP mass ratio in the epilimnion of Lake Kinneret during 1969–2001.

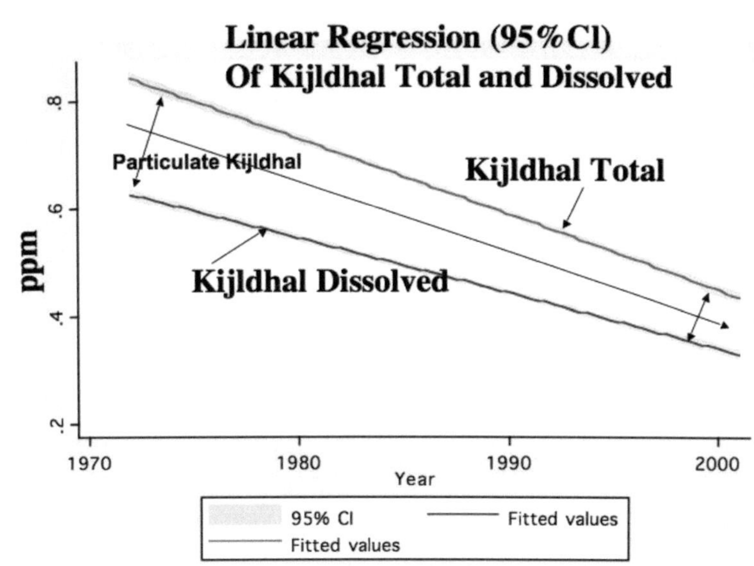

Figure 18.
Linear regression with 95% CI of temporal changes of total Kjeldahl and dissolved Kjeldahl in the epilimnion of Lake Kinneret during 1969–2001.

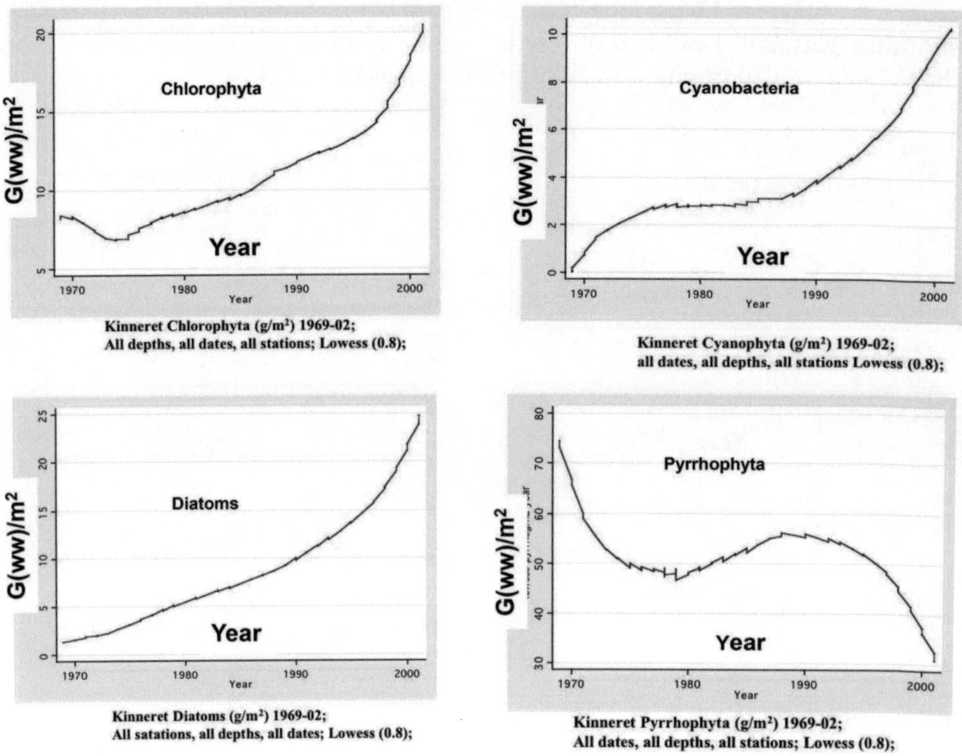

Figure 19.
Temporal trend of changes (LOWESS 0.8) of monthly means of the biomass (g(ww)/m²) of chlorophyta (upper left), cyanobacteria (upper right), diatoms (lower left), and Pyrrhophyta (lower right) in Lake Kinneret during 1969–2002.

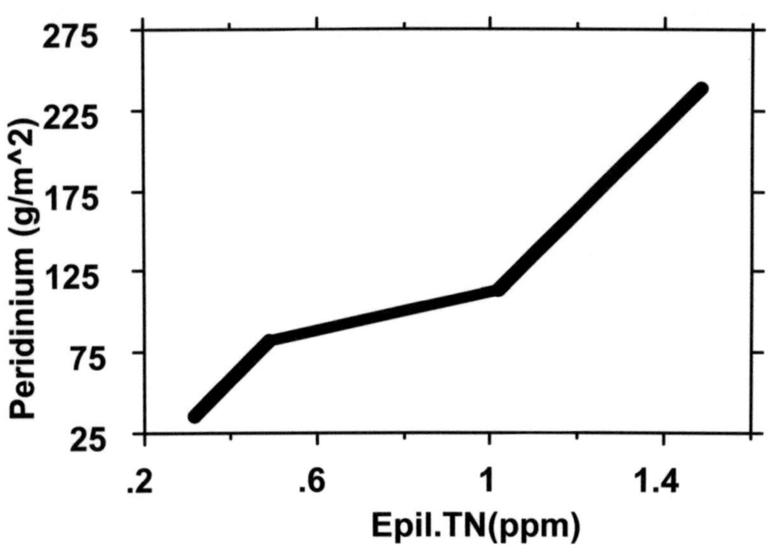

Figure 20.
Linear regression between multiannual averages of epilimnetic TN concentrations (ppm) and Peridinium biomass (g(ww)/m²) in Lake Kinneret during 1969–2001.

4. Discussion

Ecological changes of natural ecosystems were widely studied and documented. Nevertheless, cases such as Lake Kinneret-River Jordan ecosystem, which is strongly affected by anthropogenic intervention, are a typical issue being under crucial scientific and practical research. This is aimed at both prevention of natural and the human society benefits. The Kinneret-River Jordan ecosystem was under complete natural impact until the early 1930s of the previous century, while later on the "Anthropocene Era" of this system was started [5] accompanied by natural climate change. The latest periodical season (1970–2018) was especially a very sensitive factor of significant impact on this system due to two major constrains: (1) water consumption and agricultural development in the watershed and (2) lake water supply. The major changes of regional climate change and the modifications within the Lake Kinneret Limnological trait are briefly presented: elevation of air temperature and Lake Kinneret water, decline of precipitation, agricultural and hydrological management of the Hula Valley land, decline of lake water level, reduced input loads of nitrogen and phosphorus accompanied by reduction of nitrogen and a slight elevation of phosphorus in the Kinneret Epilimnion, dominance replacement of Peridinium by cyanobacteria, chlorophytes, and diatoms. The tentative objective of this paper is aimed at an answer to the question: why and how were those changes developed? It was previously documented [6] that Nitrogen sources for Lake Kinneret are mostly external and mostly effective is nitrate from the Hula Valley Peat soil degradation and fishponds and domestic sewage. Therefore, after removal of sewage and restriction of fishponds, input loads of organic nitrogen were reduced significantly and the supply of nitrate is primarily precipitation-dependent. Nevertheless, input loads of phosphorus from the watershed were reduced but P availability for the Lake Kinneret biota was slightly enhanced due to internal flux from the sediments and dust storm deposition. The dynamic changes of Phytoplankton composition in Lake Kinneret followed the nutrient alterations: the nitrogen consumer Peridinium was declined and cyanobacterial nitrogen fixers were enhanced [7].

4.1 Water level and lake shrinkage

A disputable issue was highlighted: agricultural water consumption or climate change?

Agricultural population in the "Upper Galilee" region (Upper Jordan River Watershed) was initiated during the early 1920s. From the very beginning, water supply for agricultural development was in the past and presently continues to be a major national concern. Nevertheless, the upper Jordan Watershed headwaters are the major source for the Lake Kinneret water budget but also for agricultural consumption. The multipurpose services of the Kinneret ecosystem are aimed at water supply, fishery, recreation, and tourism. Kinneret is the only natural freshwater lake in Israel, and environmental and water quality protection is essential. Moreover, water sharing is crucial and precautionary nationally and internationally guaranteed. Availability of preferential pathways in the undergrounds of the Hula Valley for subterranean water migration (obviously gravitating) is not an imaginary black hole but integrated component of the Hula Hydrological system. Evaporation is supposed to be reduced by enhancement of vegetation cover (Karacus 2019; [8–12]). The Hula Project Monitor Data Base documented annual 1376 mm with variations of 10% during the last 10 years and averaged as 1401 mm/y. This record is based on climatological parameters (Penman-Monteith equation). Multi-annual record (1960–2018) indicates no significant change in the vegetation cover in the Hula Valley region except the seasonality of crop cycling. The ET capacities were, therefore, potentially stable more or less unless climate conditions changed. That is because ET records in the Hula Valley (Hula Project–Migal Data Base) are mostly climatologically dependent (Penman-Monteith formula). Wine [13] and Wine et al. [5] documented the absence of significant climate change conditions in the Upper

Jordan watershed. Givati [1] approximated an increase of 40 mcm/y consumption of Kinneret Headwaters.

Water legislative consumption inspected by the "Water Authority" was significantly reduced from >100 to 68 mcm/y at present implementation. Confirmed information documented restriction from 200 to 68 mcm/y. Conclusively, climate change, precipitation decline, and restricted legislative allocation essentially make sense. The whole Hula Valley is under plant cover since the late 1950s of which 48×10^3 dunams are presently tree covered (groves). The Golan Heights area is tree covered (grove and vineyard) by about 74.5×10^3 dunams. Consequently, total tree covered area of the Upper Jordan Watershed Upstream during 2017 is about 123×10^3 dunams. A small quantity of water ($14–17 \times 10^6$ m^3) is also conveyed from headwater sources (Hula Valley Western Canal) to irrigate crops on the Dalton Plateau (Western part of the Upper Galilee). Water consumption for irrigation in this part of Israel is fully monitored and absolutely restricted to legislated permission.

4.2 Vegetation cover changes in the Hula Valley

Water supply for irrigation on the Golan Heights (part of the upper Jordan watershed) is precipitation-dependent. Thirty-two reservoirs (total capacity 34 mcm; 10^6 m^3) were constructed on the Golan Heights to store natural runoff water. Therefore, maximum water reduction from the Kinneret budget is 34 mcm/y. The maximum reduction of water removal from the Kinneret budget attributed to the increase in vegetation cover is worth only to 20 cm of the lake water level. Maximum storage in the Golan reservoirs occurs when regional precipitation is surplus while all other resources are plentiful and Kinneret WL is increasing anyhow.

Givati [1] documented the decline of 77 mcm/y of Dan and Banias river (major sources of River Jordan) discharges and an annual decrease of precipitations on the Golan Heights of 246 mm during 1970–2010 (Mean 6.2 mm/y).

A reasonable option for invisible water loss in the Hula Valley is the underground preferential pathways gravitating subterranean water migration. Gophen [3, 4] and Gophen et al. [2] suggested 38 mcm/y water loss during drought seasons in the Hula Valley. The Kinneret water balance management was optimized between the dependents of natural parameters (precipitations and river discharge) and human demands (agriculture and domestic). The lake water level resulted in the fluctuations caused by those fluctuating parameters. The national supply (direct pumping from the lake and northern consumption) continued until a severe long-term (5 years) drought led to insufficiency; then a desalinization solution was implemented. Throughout this long-term process, Galilee and the Golan were populated, followed by agricultural flourishing, and the lake was not devastated.

Calculated drought level using the SPI (standard precipitation index) scale during 1930–2016 [1] indicated 13 and 17 years of severe drought during 1930–1970 and 1970–2016, respectively; it also indicates climate change.

The search for well-known cases of water level (WL) decline in lakes is common among limnologists and hydrologists. Deterioration of water quality quite often by water level decline is accompanied by other factors. However, shrinkage measure of the lake is closely related to the Bathymetric features of the lake. In shallow lakes with a flat bottom surface (such as the Aral Sea and Lake Chad), a minor decline of WL exposes a vast bottom area and extreme shrinkage while in deep lakes (steep bottom surface) the opposite occurs: exceptional WL decline exposes a smaller area of the bottom surface. A WL amplitude fluctuation of 20 m (197–217 mbsl) in

Lake Kinneret during 9000 years did not threaten its existence as much as a decline of 19.5 m in Lake Sivan (Armenia). Water utilization for agriculture sometimes (Aral Sea and Lake Chad) led to a huge shrinkage in water bodies, while in other cases (Kinneret, Sivan), a much smaller shrinkage occurs. The top priority of Lake Kinneret exploitation is water supply for domestic use. Removal of water from the Kinneret budget for agricultural development north of the lake by the national authority has led to exceptional shrinkage of Lake Kinneret, which is not realistic. Insufficient replenishment of the Kinneret storage capacity by reduction in water inputs is due mostly to climate change.

5. Conclusive remarks

The impact of climate change in the Upper Jordan Watershed was precipitation and consequently river discharges and available water capacities since the mid-1980s decline [14–16]. Regional air temperature increase was accompanied by the elevation of Lake Kinneret water temperatures. The ET regime during 2005-2018 in the Hula Valley was approximately stable (± 10%). Air temperature increase did not caused significant fluctuations of the ET values due to the high density of vegetation cover which reduced soil surface warming. The changes of the ecological Lake Kinneret trait were mostly due to nutrient dynamics of limitation shift from Phosphorus to Nitrogen.

Author details

Moshe Gophen
MIGAL, Scientific Research Institute, Kiryat Shmone, Israel

*Address all correspondence to: gophen@migal.org.il

References

[1] Givati A. The impact of climate changes on water resources in the North, Water Authority. 2016 (Unpublished data)

[2] Gophen M, Meron M, Orlov-Levin V, Tsipris Y. Seasonal and spatial distribution of N & P substances in the Hula Valley (Israel) subterranean. Open Journal of Modern Hydrology. 2014;**4**:121-113. DOI: 10.4236/ojmh.2014.44012

[3] Gophen M. A climate change and water loss in the Kinneret drainage basin. Land Use Policy. 2018;**80**(2019):424-429. DOI: 10.1016/landusepol.2018.03.008

[4] Gophen M. Water-mediated nutrients export from the Hula Valley to Lake Kinneret: A review. Global Journal of Human-Social Science-H Interdisciplinary. 2018, 2018;**18**(8 Version 1.0):20-39. Online ISSN: 2249-460X & Print ISSN: 0975-587X

[5] Wine ML, Rimmer A, Laronne J. Agriculture, devcersion, and drought shrinking Galilee Sea. Science of the Total Environment. 2019;**651**(2019):70-83. DOI: 10.1016/j.scitotenv.2018.09.058

[6] Gophen M. Scientific values are crucial for Lake Kinneret management design: Test cases. Engineering Management Research. 2016;**5**(1):12-21

[7] Gophen M, Smith VH, Nishri A, Threlkeld ST. Nitrogen deficiency, phosphorus sufficiency, and the invasion of Lake Kinneret, Israel, by N_2-fixing cyanobacterium *Aphanizomenon ovalisporum*. Aquatic Sciences. 1999;**1**:1-14

[8] Bokaie M, Zarkesh MK, Arantesh MP, Hosseini A. Assessment of urban heat island based on the relationship between land surface temperature and land use/land cover in Tehran. Sustainable Cities and Society. 2016;**23**:94-104

[9] Chen XL, Zhao HM, Li PX, Yin ZY. Remote sensing image-based analysis of the relationship between urban heat island and land use/land cover changes. Remote Sensing of Environment. 2006;**104**(2):103-146

[10] Julien Y, Sonrino JA, Verhoef W. Changes in land surface temperatures and NDVI values over Europe between 1982 and 1999. Remote Sensing of Environment. 2006;**103**:43-55

[11] Kaufmann RK, Zhou L, Myneni RB, Tucker CJ, Slayback D, Shabanov NV, et al. The effect of vegetation on surface temperature a statistical analysis of NDVI and climate data. Geophysical Research Letters. 2003;**30**(22):2137

[12] Li W, Bai Y, Chen Q, Hee K, Ji X, Han C. Discrepant impacts of land use and land cover on urban heat islands: A case study of Shanghai, China. Ecological Indicators. 2014;**47**:171-178

[13] Wine ML. There are no black hole swallowing water in the Hul Valley. Landuse Policy. 2019;**84**:363-364. DOI: 10.1016/j.landusepol.2019.02.046

[14] Karakus CB. The impact of land use/land cover (LULC) changes on land surface temperature in Sivas City Center and its surroundings and assessment of Urban Heat Island. Asia-Pacific Journal of Atmospheric Sciences. Springer B.V. 2019. ISSN 1976-7633. DOI: 10.1007/s13143-019-00109-w

[15] Rimmer A, Gal G, Opher T, Leschinsky Y, Yacobi Y. Mechanisms of long-term variations in the thermal structure of Lake Kinneret. Limnology and Oceanography. 2011;**56**(3):974-988

[16] Tsipris J. ET data in the Hula Valley: Migal Hula Project Data Base. 2005-2018

Chapter 9

Rural Tourism and Territorial Development in Italy

Manuel Vaquero Piñeiro, Paola de Salvo and Francesca Giommi

Abstract

This chapter analyses the development of rural tourism in Italy over the recent decades. Since the end of the twentieth century, also due to the crisis of industrialization and the exhaustion of the factory work model, interest in "returning to the countryside" has grown. The spreading of new social and cultural trends, more sensitive to the issues of sustainability and preservation of natural wealth, has encouraged in Italy, the country of art and the beautiful landscape, a sudden increase in new tourism initiatives that are environmental friendly and revolving around quality food and wine. The "Roads of flavours" and numerous associations of small towns have therefore been created, to enhance the Italian artistic and architectural heritage. In particular, interest in Mediterranean diet has increased, as shown by the growth in tourism linked to oil. There is no lack of compatibility problems as well as difficulties in devising new cohabitation formulas, but as this work demonstrates, after the boom of the "sun and beach" tourism of the 1960s, today a far more diversified tourist panorama has emerged, thus allowing the rural world to benefit from a renewed momentum.

Keywords: Italy, tourism, land, food, development

1. Introduction

Since the ending of the twentieth century, rural Europe has been at the core of a deep social change, which saw the countryside to evolve from a simple "production landscape" into a "consumer landscape" [1]. With the spreading of the service economy, the countryside turned into a mainly residential and recreational place [2]. It therefore increased its attractiveness by becoming more accessible and more useful, thus providing rurality with a new role within the broader social context. This change is derived from a plurality of social as well as cultural processes, such as the growing mobility and integration among goods, services, people and knowledge from different territorial areas, including rural and urban areas [3], together with a healthier life style associated with a new concept of well-being. Under these circumstances, agriculture gained new significance, by becoming a priority of rural development [4]. During the last decade, indeed, numerous studies emphasized the relation of agriculture with the environment, with other economic sectors, and with society as well. As pointed out, this approach resulted from a change in the economic and social meaning of the primary productive activities, which attributed to agriculture a different and wider role than in past times.

IntechOpen

In this framework, consecutive Common Agricultural Policy (CAP) reforms progressively broadened the rationale of the rural development policies, from the simple support to the fostering of its development through the promotion of the agricultural policies, a better care and financial aid aimed at both qualifying and preserving the environment and the rural space, the improvement of the life quality, as well as a growing diversification of the rural economy. When, at the beginning of the twenty-first century, we talk about "the revenge of the countryside" [5] and of the economic value inherent in the rediscovery of the identity of the places [6], the concept of rural development was linked to the new role attributed to agriculture in relation to the environment, to the territory, and to food safety and quality. If in the second half of the last century announcing the end of peasants, erased by industrialization and the growth of the cities [7], nowadays a new scenario of rurality is emerging, so-called the postindustrial or postmodern rurality, where agricultural development requires policies based on the territory and oriented toward general development. With the progressive establishment of a multifunctional agriculture, coherent with the guidelines of the Economic Community policy [8, 9], which is able to play different roles in addition to the traditional primary one linked to the production of food and raw materials, and the growing diversification of economic activities within the rural areas, the modern agriculture has played a different role in respect to the past, having to respond to the renewed needs of the current society in terms of food safety, environmental protection, recreational needs, and, in general, the improvement of the quality of life [10]. The general objective of this work is the analysis of the relationship among local food, rural territory and tourism. Specifically, we will investigate on how tourism linked to the enhancement of food can contribute to the development of activities related to its production, linking it to new forms of tourist experience, helping to preserve and disseminate the rural culture of a territory.

As registered by the World Tourism Organization (WTO), tourism connected to food and countryside is a quickly growing segment. A high percentage of tourists does choose tourist destination on the basis of food as well as of the will to gain positive experiences by tasting local products in small locations, renowned for their quality and their strong connection to the territory. In this way, a strong relationship has been created among free time, tourism, food rediscovery, and rural areas' promotion, able to exert an influence on both demand and supply of tourist services.

Local traditional products, in Italy, represent a decisive component of the endogenous development of territorial systems (Food Clusters, Metropolitan Food Clusters), due to the significant economic, social, and tourist repercussions they can produce [11]. Thanks to typicality promotion, the productive function of the agricultural activities is integrated with new and diversified functions, among them are the environmental and territorial safeguard, the preservation of culture and rural traditions, and the creation of spaces and locations interested in new economic and social dynamics [12]. During the last years, it has been possible to observe a proliferation of initiatives aimed at matching the agricultural production with services (touristic, recreational, educational, social, and other services), also in order to intercept and satisfy new consumer segments interested in the fruition of agricultural products in the territory of their production, so to "plunge" within the local culture and to live consumer experiences as opportunities of cultural and social enrichment.

All the same, local public administrations look at rural as well as gastronomic tourism with renewed interest, also in the light of a strengthening of both local community identity and cohesion, by fostering synergies and links with other territorial economic activities (handicraft, tourism, etc.) so to favor a local endogenous development. The special focus we put, with this study, on the gastronomic products has also been encouraged and supported by the growing public care for food quality, other than by the will of promoting as well as preserving local traditions and by a more general sharing of a simpler and more natural life style. It is after all of universal acknowledgement that traditional products, as forms of expression of a territorial culture, strongly affect the social and economic development of rural territories.

2. Toward a new vision of rural development: the relationship between local gastronomy and tourism

By borrowing the definition from the World Food Travel Association, "*food tourism is the act of traveling for a taste of place in order to get a sense of place.*" The development of gastronomic tourism has prompted the interest of the scientific community that has begun to describe and theorize about this new social phenomenon. The research was recently directed to consider the culinary tourism as a new emerging tourist practice, showing a clear and steady increase in travel-oriented food [13, 14].

The scientific literature initially highlighted how local food could become a new tourist attraction [15] and how new tourist expectations could be created around it [16]. The study of gastronomic tourism has led some authors [17, 18] to highlight some important aspects for its development and its affirmation, identifying four categories that refer to structures, activities, events, and organizations [19]. Regarding the structures, the authors mainly refer to buildings (wineries, olive groves, farms, museums, taverns, etc.), land use (vineyards and olive groves), and to the itineraries (wine and olive oil roads, etc.). With reference to the activities, gastronomic tourism is linked to the methods of consumption of the product (tastings, pick your own activities, etc.) and to the experiences of education in the knowledge of food (cooking schools, visits to the places of production, etc.). As regarding the events, which are the third category identified by the authors, we can identify fairs, parties, and events linked to food. Finally, as regards the organizations, we refer to the presence of certification systems and the structures of the product and adherence to civil society associations that enhance the connection of food to the territory (Slow Food, National Identity Associations such as City of olive oil, City of wine, of pasta, of hazels, etc.).

In the last 20 years, food positioning in the tourism sector has changed profoundly and geographical destinations have recognized their gastronomic potential as an important attraction factor and as a new opportunity to position themselves in an even more increasing competitive global market. The interest in typical products and local food has thus acquired an ever-increasing importance among tourist motivations [20], thanks not only to the renewed attention to gastronomy that characterizes today's society, but also to the affirmation of a tourist offer of experience complex, based on gastronomic resources capable of involving and stimulating the sensorial and experiential component of tourism consumption. A gastronomic tourism is affirmed that is a movement of tourists who, while moving, buy and consume local food, observe and participate in the process of food production and consider it as the main motivation to move or at least as one of the most important activities that characterize the journey [17]. About 59% of Italian tourists consider the important or very important the presence of an enogastronomic offer and thematic experiences.

The evolution of the relationship between gastronomy and tourism has, however, been characterized by some important steps that have changed the interest and the involvement of tourists with respect to local food. Richards [20] in deepening the relationship between gastronomy and tourism highlights what he defines the three generations of the development of this relationship. The first attempts to bring tourism and gastronomy closer together, which is what Richards calls the first generation of this relationship, were exclusively linked to the development of tourism-related experiences by producers for consumers [21]. In this phase, tourists approach the places and other cultures through food and begin to assert what will be called gastronomic tourism [22]. This first and simple approach is overcome when food begins to be considered as a possible tourist attractor to enhance and promote within the territorial tourism development strategies. If food had always been an important element for tourism, tourists have always needed to eat, now becomes one of the main reasons for visiting destinations, and food becomes a tourist experience to be practiced. These changes reinforce what is referred to as the second generation of the relationship between gastronomy and tourism, and it is in this new framework that the first activities of co-creation of gastronomic experiences have been established for a decade and consumers/tourists acquire an important role proving to know the food and local productions as much as the producers themselves.

The greater awareness acquired by consumers in the processes of creating tourist experiences linked to the local gastronomy is legitimized by the birth of the *foodies* [23, 24], figures that will be decisive for the development of gastronomic tourism. Barr and Levy [23] are the first to use the term foodies and in their text "The Official Foodie Handbooks" define them as:

> *"a person who is very very very interested in food. Foodies are the ones talking about food, in any gathering over restaurants, recipes, radicchio ... They don't think they are being trivial-foodies consider food to be an art, on a level with painting or drama."*

In recent years, gastronomy played a central role in determining travelers' expectations and motivations. Food and local culinary specialties can be currently seen as real touristic attractions and able to move a food travelers' target or "*foodies*" [25]. In this way gastronomy, as well as representing a pleasant sensory activity, does evolve into both an attraction factor and a tourist marketing tool for destinations [26]. Food, in other words, become the unifying point between territorial authenticity and a tourist ever more interested in genuine, inclusive proposals, strictly linked to the territory he is going to visit. This association is even more valid for countries such as Italy, where the combination of agriculture, agri-food products, and preservation of the historical landscape is strong. In the Mediterranean countries (Italy, Greece, and Spain), the food and gastronomic landscapes play a fundamental role in the image construction.

The affirmation of the figure of the *foodie*, which is something more than a simple gastronomic tourist because his interest in food is part of his daily lifestyle and is not just linked to the trip [27], has certainly contributed to the affirmation of not only new tourist destinations, but also of restaurants, bars, taverns, and typical places. The growing ties between food and travel become evident also thanks to the spread of not only tourist guides and specialized publications, but also of specially dedicated websites. The growing number of *foodies* is leading to adaptation of destinations to their requests and needs. Tourist destinations become *foodscape* or places that enhance and create spaces dedicated to meeting the needs of food lovers. The *foodies* are attracted by *foodscape* not only to eat and taste local food but also to stay and visit the chosen destination [20] that becomes a gateway for the local identity of knowledge and approaches tourists and residents leading them to experience a common cultural experience. It creates a unique link between food, landscapes and cultures, stimulating an active participation of consumers in the activities of producers: tourists buy typical food products in local markets, participate in cooking classes, which use local products, choose restaurants with greater frequency, inns, and local establishments, thus determining a direct contact between those who create and those who consume food-related experiences.

Finally, the third and current generation is that linked to the enhancement of the landscapes of food, and the relationship between gastronomy and tourism is increasingly linked to the dynamics of local development and the revaluation of the landscape as an expression of all its elements, including the course of the food, the growth, and the development of *foodscape* that demonstrates this evolution. This creates spaces and places that are affected by new economic, social and tourist dynamics that satisfy new segments of consumers/tourists interested in the use and consumption of local food in the territories of production in order to "immerse themselves" in the culture of the places and to live the consumption, as cultural-enrichment opportunities, and social experiences. The *foodscape* becomes more and more a sensorial landscape, a unicum of flavors and aromas, which helps to stimulate research and the choice of a particular place to visit and which becomes an element that mainly directs the choice of the tourist [14, 28].

2.1 The role of local productions for the development of rural tourism

Local productions are part of the culture of a rural territory; they are elements of the past, expression of the traditions of a place that contribute, as argued by Dallen and Boyd [29], to a dynamic conservation of the landscape. Local products,

Figure 1.
The regional distribution of Italy's typical products. Source: our elaboration from www.politicheagricole.it.

also expression of human actions, become a product of the territory and even more an element of attractiveness and of tourist interest (**Figure 1**).

In the twenty-first century, scientific innovation and globalization have intensified the interest in food, offering a wide variety of food products, encouraging the emergence of alternative food movements, supporting a new food culture that has also enhanced the success of a gastronomic tourism. In this historical moment, food is the subject of discourses and reflections of critical and interpretive positions and is an expression of values, problems, and priorities of today's society. It has already been underlined how the discourses around food concern food safety, health, the ethical treatment of animals, the industrial impact, food production, the development of alternative food strategies, the symbolic dimension of food practices, and the socioeconomic potential that they have for local communities. So, there are numerous discourses that can be built around food: it is normally the consumerist discourse that characterizes the reflections on food, because its consumption is fundamental for life, but in the global context, we are witnessing the affirmation of new tensions and comparisons that lead to new paradigms that enhance the ethical and social aspects of food choices and promote local foods. Portman [30] argues that the defense of local food triggers discourses and new interests that rest on the awareness that food choices today no longer respond exclusively to economic indications, but increasingly reflect precise political and moral choices oriented to the issues of quality and sustainability. In fact, a new model of development has been affirmed within which local food is associated with numerous benefits, among which the increase farm incomes, greater social vitality, territorial regeneration activities, the enhancement and protection of traditional activities, and also the development and promotion of a gastronomic tourism that enhances the link between food and territory and between local cuisine and culture from which it originates [31]. In recent years, scientific literature has shown that local food, also considered a cultural and social capital [32, 33], can contribute not only to socioeconomic well-being of rural areas [34, 35] but also to their tourism development [20, 36, 37].

The value of food not only goes well beyond being a simple means of livelihood but can also contribute to the improvement of the tourist experience. The consumption of local food offers in fact a number of sensorial tourism experiences that link the local gastronomy to the journey and the discovery of the places [38]. Promotion of typical products through the creation of territorial as well as collective brands, hinged on the memory of places and communities, is able to trigger endogenous, integrated, and sustainable rural development paths. Therefore, both the promotion of traditional products and the development of a gastronomic tourism may represent precious opportunities in contrasting the current economic crisis. Especially in the less favored and marginal rural areas, far from the modernization process, a self-centered rural development model can make a root, characterized by the use and reproduction of experiences and knowledge locally developed to convert local resources into quality agri-food products while able, at the same time, to preserve the local elements, which are its foundation.

Within the frame of this new model, typical products become a resource capable of adding value to the development of smaller areas because they are able to integrate and enhance different territorial resources [34, 35], therefore responding to changes in the style of consumption of postmodern tourists. How to promote different territorial resources depends on the actors involved and on the strategies they decide to pursue. Traditional products, therefore, represent a potential resource for the local community and around them revolve several inclusive dynamics as well as projects of collective development. Typicality does not exclusively rest on the features of the productive process, but also on relations among the actors of the territorial system, which provide the typical product with a collective dimension.

The collective nature of typical products, together with their ability to promote territorial identity, quality and culture, is bringing about the shaping of new social networks and able to steer the choices regarding local development in a direction more sensitive to issues such as the development sustainability, the communities' quality life, and the enhancement of the territorial identities [39].

2.2 Tasting of typical products: the food trail

Other than for art, Italy is worldwide identified as the country where you eat and drink well. Words as "pizza," "pasta," "espresso," and "cappuccino" have conferred to food a universal dimension, thus becoming brands of a gastronomic globalization. Italy is actually one among the countries where the food and wine offer is wider and more diversified. With its 863 appellations (**Figure 2**)—299 IG food (35%), 526 IG wine geographical indications (61%), and 38 IG spirits (4%)—Italy comes before France (764) and Spain (358). The most important years in the creation of geographical indications of origin go from 1996 to 2012. Distributed by regions, agri-food IGs predominate in Emilia Romagna (45), Veneto (38), and Lombardy (36), while wine appellations prevail in Piedmont (59) and Tuscany (58). Furthermore, in Italy, there are 172,688 restaurants of which are 586 of excellence, 11,632 farms with restaurants, 18,632 farms with accommodation, 170 wine routes

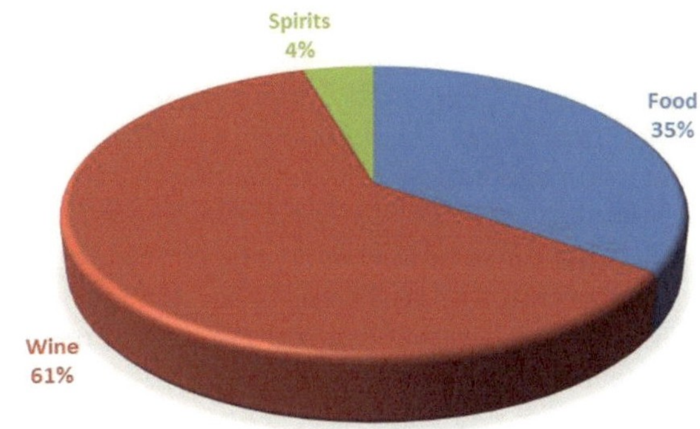

Figure 2.
Geographical indications of Italy's food (2019). Source: our elaboration from www.qualigeo.eu/statistiche-eu-dop-igp-stg/.

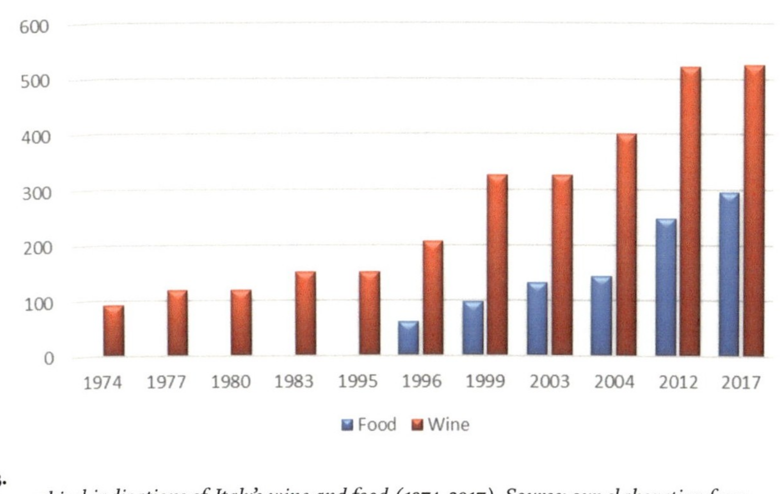

Figure 3.
Trend geographical indications of Italy's wine and food (1974-2017). Source: our elaboration from www.qualigeo.eu/statistiche-eu-dop-igp-stg/.

and flavors, and 99 museums of taste (reports on gastronomic tourism, data from 2016 to 2017) (**Figure 3**).

These few numbers help to understand the articulation of the gastronomic tourism economic background, which merges together countryside and city, family kitchen and starred cuisine, large markets and small shops, private companies, and public policies. In the first year of life, the *Fabbrica Italiana Contadina* (FICO), the large thematic area inaugurated in Bologna, had 1 million visitors, and the turnover was 50 million euros. Another Italian event related to food and that attracts hundreds of people is the Vinitaly organized every year in Verona. In the last editions dedicated to showcase Italian and world wine, there were more than 200,000 visitors, demonstrating the economic and media force of great food-related events in contemporary society [40].

Special attention deserves the routes of wine, oil, pasta and flavors because they are networks spread all over the Italian peninsula by involving a large number of small municipalities. The routes of flavors are a tool, which offers to a vast part of the Italian peninsula the opportunity of an evolution oriented toward both the upgrading and the development of the territory. This process is important to promote the development of less economically advanced areas, such as mountain areas currently experiencing a loss of population as well as of economic activities. In these areas, in fact, most of the typical products are produced (3200 local productions come from the Italian mountain areas), they in these territories, less favored and tendentially excluded from modernization processes, become a resource capable of giving value to their development by integrating and enhancing the different territorial resources. Moreover, it is unanimously recognized that typical products, as a form of expression of the culture of a territory, greatly influence the social and economic development of local rural territories, in particular through the achievement of the following socioeconomic benefits: the increase in income from agricultural enterprises, greater social vivacity, a territorial regeneration, through the valorization and conservation of traditional activities, and finally also the development of a food and wine tourism that can contribute to improving the economic sustainability of these territories.

On the backdrop of a global redefinition of economic balances, the routes of typical products—much of them located also in marginal territories and mountains areas—operate at different levels, including the promotion of tourist systems and thematic routes that are able to bring back vitality to marginalized areas faraway from major routes and rail links. Geographical marginality can, therefore, become a useful resource, and the routes represent an efficient tool of its promotion and image communication. They include a very articulate set of initiatives, which involves a wide number of public and private actors variously participating a network, which is going to become, thought in a way not always systematically integrated, the engine of new forms of local development among shapes, structures, and geographies of a constantly changing territory.

The innovative dimension of the roads' governance is based upon the necessity of building strategic alliances, which demand territories to establish new relations reshaped according to common interests and objectives, and it can now operate through a multitude of instruments of various kinds, such as strategic plans and programs, projects of territorial, urban, environmental and tourism marketing, integrated projects, etc. These tools represent an innovative approach for the governance of territorial development and transformation processes in terms of environmental and socioeconomic sustainability, not last thanks to the use of adequate information systems, the formation of consent, and the participatory construction of decision-making processes.

A route of typical products originates from the will and the union of the strengths of numerous subjects, linked together by an effort of engagement and networking led within the frame of a territory. It is possible to identify its main components by analyzing the product (wine, oil, pasta, cheese, rice, etc.), the territory (countryside, burgs), the ecosystem (landscape and environment quality enhancement), and the key players: producers, tour operators, associations, and institutions.

The routes, therefore, represent an example of public-private collaboration that aims at channeling, through thematic routes, various subjects of a specific territory, which compose a supply chain around a typical product, thus integrating the product knowledge and tasting as well as the tourist use of territory in a coherent meta-market. In fact, the establishment of the routes determines the implementation of a tourism development strategy especially addressed to more fragile territories, such as the rural areas, where the main tourist attraction is the presence of an intangible heritage that must be enhanced through experiential, sustainable, and quality development models.

More specifically, the Italian legislator has enacted a specific national law for the wine routes whose aims, however, can be extended to other typical products such as oil, cheese, etc. The roads of wine are indeed governed by the law of the 27th July 1999, n. 268, where, in Article 1, they are defined: "(...) paths marked and advertised with appropriate signs, along which gather natural, cultural and environmental values, vineyards and cellars of single or associated farms open to the public; they represent an instrument through which the territories and the relative productions can be disseminated, commercialized, and enjoyed in form of tourist offer." Main goal of the law is to promote territories with a wine vocation, with particular reference to places with high-quality productions. National legislation highlights how a route can be considered as an integrated system of both territorial and tourism offers organized along a route characterized by places of historical, artistic and environmental interest, flanked at the same time by a series of structures of reception, promotion and marketing of local products. Definition in art.1 of the 1999 national law reflects an approach consistent with the new multifunctionality of the rural world. This regulatory intervention provides the necessary prerequisites in order to increase the competitiveness of the territorial systems and to contribute to the formation of the roads themselves, by aiming at creating effective networks characterized by a bottom-up approach, which require active partnership as well as planning skills on the part of the public and private operators belonging to both the chain of typical products and the tourism sector (**Figure 4**).

Wine tourism turnover in Italy ranges between 2.5 and 3.5 billion euros as a result of a tourist movement involving about 4 million people. A prime example of transformations connected to the gastronomic development is the small Tuscan town of Montalcino where, between 2011 and 2016, tourist arrivals related to wine increased by 125% (from 5000 to 24,000), thus favoring a multiplication of hotel facilities (from 14 to 50) and beds (from 78 to 620). Montalcino, which takes advantage of the excellent international image enjoyed by Tuscany, is not a lonely example of growth. Among other areas in Italy that in recent years have experienced a radical change in their image are Conegliano (Veneto), Oltrepò Pavese (Piedmont), the Langhe (Piedmont), and Montefalco (Umbria). These places are a clear demonstration of changes that have taken place in Italian tourism over the last decades when, under the effects of the industrialization crisis and of the growing criticism about the massive construction of the coasts, the tourism model moved from a mass "sun and beach" tourist offer toward a more diversified idea of how to enjoy a free time. In the light of this new tourist demand, related to the attractiveness of the campaign and of the consumption of typical products, producers also were induced to adapt

Figure 4.
Wine routes in Italy (2019). Source: our elaboration.

their structures from simple places of production into suitable structures for tourist accommodation.

Some figures can help to clarify. In Italy, there are today more than 1200 cellars equipped to receive tourists and the cellars that can be visited are over 12,000. During individual visits and wine tours, they offer tastings, visits to the vineyards, and other recreational and educational activities. Estimating the economic impact of food and beverage tourism is, at best, very difficult. In an approximate way, knowing that this industry is characterized by a particular type of tourists—the food travelers—which escapes easy definitions and estimates suggest that the wine tourists who visit the cellars bring about a 31.35% increase in turnover, an increase that should be distributed through the entire tourist supply chain (hotels, restaurants). These percentages recall those of the WFTA, according to which approximately 25% of visitor spending can be attributed to food and drink while traveling. This change took place almost naturally in extra-European countries such as Australia or New Zealand, where the wine tourism sector grew together with viniculture, while in Italy and in the European countries in general, the transformation of wineries in tourist destinations has induced a business change in order to adapt buildings and rural settings to visiting activities. A change was also required in terms of offering the visitor the image of a sustainable production, attentive to the preservation of nature and to the application of the most modern technologies for recycling and saving energy. In this sense, wine tourism at the beginning of the twenty-first century is a perfect example of contemporary society trends.

DOI: http://dx.doi.org/10.5772/intechopen.85774

3. Landscapes and cultural heritage, a case study: the olive oil tourism in Italy

Olive tree is the most typical Mediterranean plant and olive oil, other than the symbol of a millenary culture, and is considered since the twentieth century as a fundamental food for the health of people (Mediterranean diet). Moreover, olive tree and oil are two clear examples of environmental and social sustainability. Given this cultural value, in recent years, following the wine model, tourism too has begun to look with interest at the attractiveness of the olive-covered landscapes as well as to the possibility to be acquainted with the olive oil production techniques. A specific tourist offer has, therefore, been created, which is strongly linked to both the territory and the history of the Mediterranean countries.

There are numerous studies that highlight a relationship between tourism and olive oil; initially the international literature, in particular in Australia, has highlighted how oil tourism is essentially an expression of the wider phenomenon of rural tourism [40, 41]. Furthermore, the study of tourism related to olive oil has been investigated in Europe, especially in Italy and Spain [42–45] where the researches have not only enhanced the rural aspect but also highlighted how the interest in olive oil can give value to the tourist image of a destination becoming a new narrative of the territory that underlines the link between landscapes, culture, typical products, and tourism [46]. Oil tourism does not just become a further expression of the activities linked to rural tourism and to the agri-tourism, but also to cultural tourism and for its organoleptic qualities, that of health [47]. The results of the first studies on oil tourism conducted by Ruiz Guerra [47] show how initially, the interest of tourists in the olive oil industry was linked above all to the sale of products connected to it, which then activated new forms of economy for the territory, as had already happened for wine tourism [48, 49]. However, the researches that highlight the interest of tourists increase not only for oil production, but also for production sites and landscapes [40]. Furthermore, oil tourism refers not only to representational images and to visual perceptions of places, but also to a perception relative to the other senses. In fact, the results of a recent research [39] highlight how tourists show a particular interest in visiting the mills that become a new resource for tourism development linked to the world of olive oil.

The implants for the production of oil, as well as being the places that are characterized by the proposal of emotional content and *sense making*, constitute a concrete case to be placed in the broader concept of heritage cultural [50]. In fact, oil mills are to be considered a testimony to the time of the oil production processes. As a result of the standardization imposed by industrialization, oil mills handed down the specificity of the individual territories, for example, thinking to the high historical value of the mills in the grottoes existing in the Italian region of Puglia. The historical oil mills can thus constitute fundamental architectures for the enhancement of the territory, because they allow the visitor to get closer to realities that bear witness to the knowledge and the productive know-how of a community, the relations with the territory where they insist, contributing with their presence and activities to an economic and social enrichment.

With this in mind, buildings that still retain particular architectural values must be protected and enhanced through initiatives that support the agricultural production of services (tourism, recreational, educational, and social). New forms of use of the mills are established in which tourism activity is taking a leading role, as demonstrated by the results of the research that will be presented in the following paragraphs. Olive oil tourism is certainly a recent phenomenon of tourism; it is a matter of following in the footsteps of wine tourism that has managed to place the cellars (new and old) at the center of public attention, attracted now by the valuable historical buildings now from the shapes created by modern architects [51, 52]. For olive oil tourism, the experience of wine tourism should offer points of comparison to exploit the mill in that building and not only as a working space. Next to the conventional museums (Olive oil Museums), the mills have the characteristics to transform themselves into real museums aimed at describing to the public all aspects related to the production of oil, fundamental aspect that distinguishes the needs of the new tourist related not only to the dimension of seeing but also of learning and knowledge. One could speak of a museum with a high didactic impact where to combine the value of olive oil with the weight of history and tradition.

In this new context, the dynamics of tourism, linked to the characteristics of the tourist experience, also applies to the tourism of the olive oil and to the possible and manifold offers that the tourist system connects to it. The activities that may be linked to the development of olive oil tourism involve several subjects, primarily tourists who have the opportunity to know the territory through a typical product and the tourist experiences connected to it, the producers that can enhance their business with a new way of marketing their product, and finally the tourism system that has the opportunity to redevelop and characterize its offer with specially prepared tourist itineraries with the involvement of the hospitality and catering sectors. Although a part of the literature [43] argues that the tourist itineraries linked to oil do not have a great development, in Italy, there are many experiences of territories that have activated paths of knowledge of the oil, made by both producers and institutions, which present oil as a qualifying element of the territorial offer. For example, the regional route of the POD Umbrian olive oil, whose characteristics will be presented later, qualifies as a possibility of knowledge, promotion, marketing, and enhancement of the oil throughout the territory involved in the itinerary.

The road of typical products thus becomes an instrument for the promotion and diffusion of local culture and traditions linked to oil [53]. In Italy, these roads are highly specialized and concern single local productions, wine, oil, cheese, apples, etc. But in some territories, a different orientation is starting to prevail, perceptible by the same road names, more generally defined roads of flavors, where the degree of specialization decreases, highlighting territorial experiences of itineraries that include wine, olive oil and other quality agri-food products. The presence of tourist itineraries on the olive oil is certainly one of the important tools of promotion and local valorization, along the routes the tourist has the opportunity not only to meet points of direct sale of the oil, to make visits to the mills, to taste the oil of olive, but also to observe the landscape in which the itinerary develops. In this context, olive oil tourism can help to activate development processes aimed at reconstructing the local cultural identity understood as the identification, protection, and enhancement of all the factors that contribute to creating the specificity and uniqueness of the places and the environment. There are tourists who seek and ask more and more landscape, more and more nature, and desire a deep and direct contact with the values of rural culture; the heritage inherited from the past becomes an element of interest and also of tourist interest. The olive oil, the rediscovery of the values and traditions linked to it, therefore represents the possibility of development of the territories, especially those of a rural nature, and can become the fulcrum around which to propose and/or redesign cultural tourism offers [54, 55]. Olive oil in some territorial contexts is a heritage in which a community identifies itself and makes the place in which it presents unique and can develop around itself interesting experiential tourist practices that contribute to making the tourist feel, even if temporarily, part of the history of a place. Olive oil tourism can also offer interesting answers to the question of authenticity that characterizes important and growing segments of tourist demand [56], proposing tourist experiences that trigger a knowledge and a comparison between the tourist and the identity of the territory.

In this framework, the University of Perugia carried out a specific research devoted to this innovative tourism custom by relating it with a significant initiative aimed at the promotion of olive oil, named Frantoi Aperti (Opens Oil mills), which takes place in Umbria. This research, titled "Oil tourism as a tool for enhancing the action of rural territories: the case of the PDO Umbria Oil Road"represents in Italy the first attempt of reflection and study on the theme of olive oil tourism. The main objective of the research was to understand how innovative initiatives to promote the territory as *Frantoi Aperti* (Opens Oil mills) can identify in the oil product a possible driver able to promote the development of tourist flows related to the environment, the culture and the discovery of rural areas. With reference to the methodology adopted, the research activity has contemplated the administration of a questionnaire to tourists present during the six weekends of the 2013 edition of *Frantoi Aperti*. There were collected 228 questionnaires aimed at understanding the profile of the tourist who participates in *Frantoi Aperti* investigating the reasons for his visit. The investigation and survey activities were carried out in the municipalities participating at the event "Frantoi Aperti" in 2013, which involved 30 mills with guided tours to the oil plants in processing, to the new oil tasting, and to other regional products and activities.

The exploratory survey was carried out on a sample that is not statistically representative, however, able to elaborate interesting elements of reflection about the links among territory, oil and consumption styles that could be analyzed in order to provide useful information to administrators and operators of the territory and able not only to increase but also to enhance and "spread" the quality of the initiative and therefore also its tourist visibility. The administered questionnaire provided for four different thematic sections includes: profile of the interviewed tourist; permanence and hospitality: in the second area, it is represented by the duration of the visit and the type of accommodation chosen; the motivation that pushes the tourist to participate in the Frantoi Aperti event and its level of satisfaction regarding the organization and reception and hospitality of the territory; and styles of consumption and oil product: the last section of the questionnaire was dedicated to the behavior and daily habits of food consumption of the tourist.

Furthermore, the research activity has provided the creation of in-depth interviews to a representation of privileged stakeholders of the territory (local authorities, entrepreneurs, producers, bloggers, etc.) in order to understand the motivations that led them to join "the Frantoi Aperti," the advantages and opportunities of their participation as well as their contribution to the realization of the event. The stakeholders interviewed highlighted how the increase in visibility and knowledge of both the territory and its companies is the main reason behind the adhesion to "Frantoi Aperti" on the part of local authorities, producers, and owners of accommodation facilities, who evaluate positively the initiative and consider it a possible instrument, which could contribute to the promotion and de-seasonalization of the territories involved.

This research allows to observe that a specific profile of the olive oil tourist begins to take shape: the empirical survey emphasizes as this new figure considers the interest in the oil product in itself as one of the main travel motivations and as a tourist experience to be practiced. The olive oil tourist is also interested in the knowledge of food landscape in which the relationship between gastronomy and tourism is linked to the enhancement of the landscape as an expression of all its identifying elements including food. The food and culture of the territory connected to it thus form an ideal harmony for tourists who express their interest not only for the gastronomic product itself, but also for all the aspects that characterize the landscape, its history, and its traditions [20]. Attention to food is increasingly characterized by sociocultural, political, and economic elements that are contributing to the affirmation of the concept of food of the territory, foodscape, not only as a vehicle of tradition and memory but also as a food landscape, which represents not only the complexity of production, but also of the conservation and enhancement of local productions [50]. In conclusion, olive oil tourism can contribute to the local development of rural territories with the networking of local resources and skills, through a participatory project that activates sustainable processes of value creation not only for tourists but also for the local community.

4. Conclusions

This study demonstrates the significance assumed over the last decades by the world revolving around food. Food highlights how feeding choices, in the current society, not only derived from economic conditions, but also that reflect political and social choices, are oriented by themes such as quality, authenticity, as well as sustainability. Relationship between food and territory has developed very quickly, and this new food culture has also enhanced the emergence of gastronomic tourism. Territories are today very aware of the significance of their gastronomic potential, primarily based on tourism: tourists prefer to buy and consume local food while at the same time observing and participating in its production process, by increasingly considering it as the main reason for moving or, at least, as one of the most important activities that characterize their trip.

A new tourist demand has emerged, which identifies in the typical product an element capable of combining the authenticity and specificity of a territory with its knowledge and the experience of its tasting. The tasting of the gastronomy and the knowledge of the production processes of local agri-food typicality has started indeed being considered tourist practices that allow travelers to get in touch with a heritage consisting in local history, typicality, lifestyles, and traditions.

From this work, we also derive important considerations regarding the relationship among food, rural areas, and tourism: promotion of local food can contribute to rural area development and socioeconomic well-being also through the growth of a gastronomic tourism hinged on the link with the territory, on local cuisine, as well as on the culture from which it originates.

Furthermore, this chapter stresses the value of resources and tourist services, which are developing around local products, especially promotion and knowledge paths such as the roads of wine, oil and flavors, as well as the installation of dedicated museums and the organization of festivals and events. Italian territories are increasingly characterized as foodscape, and the relationship between gastronomy and tourism is increasingly tied to the dynamics of local development as well as to the enhancement of landscape as an expression of all its identity elements, including food.

Food and culture of the territory connected to it thus form an ideal harmony for tourists and for the inhabitants who express their interest not only for the gastronomic product itself, but also for all the aspects that characterize the landscape, its history, and the cultural base. The attention to food is increasingly characterized by sociocultural, political, and economic elements that are contributing to the affirmation of the concept of food districts, as a vehicle not only of tradition and memory but also of agricultural landscapes that represent not only the complexity of production, but also the conservation and enhancement of local productions, especially in rural areas. The protection of local food production can thus contribute to the socioeconomic development of rural territories by networking the resources and skills present in them, through strategic actions that activate

DOI: http://dx.doi.org/10.5772/intechopen.85774

sustainable processes of value creation not only for tourists but also for of the local community.

Food and culture of a territory represent an ideal balance for both its tourists and inhabitants who are interested not only in the gastronomic product itself, but also in its relation with landscape, including landscape history and cultural heritage. Attention for food is increasingly marked by sociocultural, political as well as economic elements, all contributing to designing the new concept of "food districts," meant as a vehicle of tradition and memory. Especially, these new trends emphasized the significance of agricultural landscapes able to convey the idea of the complexity of the production process, along with those of preservation and enhancement of local productions, especially in rural areas. Safeguarding local food production can thus contribute to the socioeconomic development of rural territories by networking their resources and skills through the implementation of strategic actions able to activate sustainable value chain interesting not only for tourists but also for the local community.

Author details

Manuel Vaquero Piñeiro, Paola de Salvo* and Francesca Giommi
Department of Political Science, University of Perugia, Italy

*Address all correspondence to: paola.desalvo@unipg.it

References

[1] Cloke P. Conceptualizing rurality. In: Cloke P, Marsden T, Patrick Mooney P, editors. Handbook of Rural Studies. London: Sage Pubblication; 2006. pp. 18-28. DOI: 10.4135/9781848608

[2] Woods M. Rural geography: Blurring boundaries and making connections. Progress in Human Geography. 2009;**33**:849-858. DOI: 10.1177/0309132508105001

[3] Hedberg C, do Carmo RM. Translocal ruralism: Mobility and connectivity in european rural spaces. In: Hedberg C, do Carmo RM, editors. Translocal Ruralism Mobility and Connectivity in European Rural Spaces. Dordrecht: Springer; 2012. pp. 1-9. DOI: 10.1007/978-94-007-2315-31

[4] Niska MH, Vesala T, Vesala KM. Peasantry and entrepreneurship as frames for farming: Reflections on Farmers' values and agricultural policy discourses. Sociologia Ruralis. 2012;**52**(4):453-469. DOI: 10.1111/j.1467-9523.2012.00572.x

[5] Barberi C. La Rivincita Delle Campagne. Roma: Donzelli; 2009

[6] Becattini G. La Coscienza Dei Luoghi. Roma: Donzelli; 2015

[7] Mendras H. La Fin Des Paysans-suivid'uneréflexion Sur la Fin Des Paysans Vingtans après. Paris: Actessud; 1984

[8] Nazzaro C. Sviluppo Rurale, multifunzionalità e Diversificazione in Agricoltura. Nuovi Percorsi di Creazione di Valore per le Aziende Agricole Delle Aree Interne del Mezzogiorno d'Italia. Milano: Franco Angeli; 2008

[9] Abler D. Multifunctionality, agricultural policy, and environmental policy. Agricultural and Resource Economics Review. 2004;**33**(1):8-17

[10] De Castro P. L'agricoltura Europea e le Nuove Sfide Globali: Instant Book. Roma: Donzelli; 2010

[11] Timothy DJ, Boyd SW. Heritage Tourism. Prentice Hall: Harlow; 2003

[12] Belletti G, Turismo BG. Turismo, ruralità e sostenibilità attraverso l'analisi delle configurazioni turistiche. In: Pacciani A, editor. Differenziazione e sentieri di sviluppo in Toscanaruralità e sostenibilità Attraverso l'analisi Delle Configurazioni Turistiche. Milano: Franco Angeli; 2011

[13] Okumus B, Okumus F, McKercher B.Incorporating local and international cuisines in the marketing of tourism destinations: The cases of Hong Kong and Turkey. Tourism Management. 2007;**28**(1):253-261. DOI: 10.1016/j.tourman.2005.12.020

[14] Getz D, Brown G. Critical success factors for wine tourism regions: A demand analysis. Tourism Management. 2006;**27**(1):146-158. DOI: 10.1016/j.tourman.2004.08.002

[15] Higgins DM. Brands. Geographical Origin and the Global Economy: A History from the Nineteenth Century to the Present. Cambridge: Cambridge University Press; 2018. DOI: 10.1017/9781139507059.001

[16] Kivela J, Crotts JC. Tourism and gastronomy: Gastronomy's influence on how tourists experience a destination. Journal of Hospitality & Tourism Research. 2006;**30**(3):354-377. DOI: 10.1177/1096348006286797

[17] Ignatov E, Smith S. Segmenting Canadian culinary tourists. Current Issues in Tourism. 2006;**9**(3):235-255. DOI: 10.2167/cit/229.0

[18] Smith SL, Xiao H. Culinary tourism supply chains: A preliminary examination. Journal of Travel Research. 2008;**46**(3):289-299. DOI: 10.1177/0047287506303981

[19] Horng JS, Tsai CTS. Culinary tourism strategic development: An Asia-Pacific perspective. International Journal of Tourism Research. 2012;**14**(1):40-55. DOI: 10.1002/jtr.834

[20] Richards G. Evolving gastronomic experiences: From food to foodies to foodscapes. Journal of Gastronomy and Tourism. 2015;**1**(1):5-17. DOI: 10.3727/216929715X14298190828796

[21] Boswijk A, Thijssen T, Peelen E. The Experience Economy: A New Perspective. Amsterdam: Pearson Education; 2007

[22] Long L. Culinary Tourism: Exploring the Other through Food. Lexington: The University Press of Kentucky; 2004. pp. 20-21

[23] Barr A, Levy P. The Official Foodie Handbook. Arbor House Publishing Company; 1984

[24] Carter A. Review of the official foodie handbook. London Review of Books. 1985;7(24):22-23

[25] Fox R. Reinventing the gastronomic identity of Croatian tourist destinations. International Journal of Hospitality Management. 2007;**26**(3):546-559. DOI: 10.1016/j.ijhm.2006.03.001

[26] Folgado JA, Hernández JM, Campón AM. El turismo gastronómico como atractivo innovadordel destino: un estudio empírico sobre rutas gastronómicas españolas. In: Comunication presented at XIII Seminary Luso-Espanhol de Economía Empresarial, held at Universidad de Évora; 24-25 Novembre. 2011

[27] Green E. A Study of Travelers' Foodie Activity Dimensions, Demographic Characteristics, and Trip Behaviors; 2013. http://thescholarship.ecu.edu/handle/10342/1815

[28] Urry J. Consuming the planet to excess. Theory, Culture & Society. 2010;**27**(2-3):191-212. DOI: 10.1177/0263276409355999

[29] Dallen T, Boyd SW. Heritage e Turismo. Milano: Hoepli; 2007

[30] Portman A. Mother nature has it right: Local food advocacy and the appeal to the "natural." Ethics and the Environment. 2014;**19**(1):1-30. DOI: 10.2979/ethicsenviro.19.1.1

[31] López-Guzmán T, Sánchez-Cañizares S. Culinary tourism in Córdoba (Spain). British Food Journal. 2012;**114**(2):168-179

[32] Miele M. Consumption culture: The case of food. In: Cloke P, Marsden T, Mooney P, editors. Handbook of Rural Studies. 2006:344-354. DOI: 10.4135/9781848608016.n24

[33] Higgins-Desbiolles F. More than an "industry": The forgotten power of tourism as a social force. Tourism Management. 2006;**27**(6):1192-1208. DOI: 10.1016/j.tourman.2005.05.020

[34] Marsden T, Banks J, Bristow G. Food supply chain approaches: Exploring their role in rural development. Sociologia Ruralis. 2000;**40**(4):424-439. DOI: 10.1111/1467-9523.00158

[35] Brunori G, Rossi A. Synergy and coherence through collective action: Some insights from wine routes in Tuscany. Sociologia Ruralis. 2000;**40**(4):409-423. DOI: 10.1111/1467-9523.00157

[36] Hjalager AM, Richards G, editors. Tourism and Gastronomy.

London: Routledge; 2002. DOI: 10.4324/9780203218617

[37] Wolf E. Culinary Tourism: The Hidden Harvest: A Dozen Hot and Fresh Reasons how Culinary Tourism Creates Economic and Community Development. Kendall/Hunt; 2006

[38] Chang RC, Kivela J, Mak AH. Attributes that influence the evaluation of travel dining experience: When east meets west. Tourism Management. 2011;**32**(2):307-316. DOI: 10.1016/j.tourman.2010.02.009

[39] Calzati V, de Salvo P. Il ruolo degli eventi gastronomici nella promozione e valorizzazione dei territori rurali. Il Caso di Frantoi Aperti in Umbria. Milano: Franco Angeli; 2017

[40] Cavicchi A, Santini C, editors. Food and Win Events in Europe. A Stakeholder Approach. Routledege: London-New York; 2014

[41] Alonso AD, Northcote J. The development of olive tourism in Western Australia: A case study of an emerging tourism industry. International Journal of Tourism Research. 2010;**12**:696-708

[42] Northcote J, Alonso AD. Factors underlying farm diversification: The case of Western Australia's olive farmers. Agriculture and Human Values. 2011;**28**(2):237-246

[43] de Salvo P, Hernández Mogollón JM, Di Clemente E, Calzati V. Territory, tourism and local products. The extra virgin oil's enhancement and promotion: A benchmarking Italy-Spain. Tourism and Hospitality Management. 2013;**19**(1):23-34

[44] Murgado EM. Turning food into a gastronomic experience: Olive oil tourism. Options Mediterranéennes. 2013;**106**:97-109

[45] Millán Vázquez de la Torre G, Arjona JM, Amador L. A new market segment for olive oil: Olive oil tourism in the south of Spain. Agricultural Sciences. 2014;**5**(3):179-185

[46] Campón-Cerro AM, Di-Clemente E, Hernández-Mogollón JM, De Salvo P, Calzati V. Olive oil tourism in southern Europe: Proposals for tourism development of olive grove rural areas. Turismo & Desenvolvimento. 2014;**21**(22):63-73

[47] López-Guzmán T, Cañero Morales PM, Moral Cuadra S, Orgaz-Agüera F. An exploratory study of olive tourism consumers. Tourism and Hospitality Management. 2016;**22**(1):57-68

[48] Ruiz Guerra I. Análisis Cuantitativo y Cualitativo del Significado del Aceite de Oliva. Una aproximacióndesdeel Punto de Vista del Consumidor. Granada: University of Granada; 2010

[49] Mitchell R, Hall CM. Wine tourism research: The state of play. Tourism Review International. 2006;**9**(4):307-332. DOI: 10.3727/154427206776330535

[50] Alebaki M, Iakovidou O. Segmenting the Greek wine tourism market using a motivational approach. New Medit. 2010;**9**(4):31-40

[51] Harvey M, Frost W, White L. Exploring wine and identity. In: Harvey M, White L, Frost W, editors. Wine and Identity. Branding, Heritage, Terroir. London-New York: Routledge; 2014. pp. 1-15

[52] Vaquero Piñeiro M. El turismo enológico en Italia: Origen y desarrollo. RIVAR. Revista Iberoamericana de Viticultura. Agroindustria y Ruralidad. 2015;**5**(2):120-140

[53] Danielmeier T. Winery architecture. Creating a sense of place. In: Harvey M, White L, Frost W, editors. Wine and Identity. Branding, Heritage, Terroir. London and New York: Routledge; 2014. pp. 230-242

[54] de Salvo P. L'olio come risorsa culturale e turistica: Il Caso della Strada regionale dell'olio Dop Umbria. In: Calzati V, de Salvo P, editors. Le Strategie per Una Valorizzazione Sostenibile del Territorio. Milano: Franco Angeli; 2012. pp. 75-90

[55] Maggiore G, Resciniti R. Experiential marketing for value creation in cultural heritage management. In: Morvillo A, editor. Advances in Tourism Studies. Vol. 1. Milano: McGraw-Hill; 2012. pp. 45-69

[56] Simeon MI, Buonincontri P. A model for planning and managing experiential cultural products. In: Morvillo A, editor. Advances in Tourism Studies. Vol. 1. Milano: McGraw-Hill; 2012. pp. 78-91

Chapter 10

Horizontal and Vertical Archipelagoes of Agriculture and Rural Development in the Andean Realm

Christoph Stadel

Abstract

The tropical Andes offer a unique mosaic of physical and human environments. Since the pioneering field research of Alexander von Humboldt over 200 years ago, the Andean realm has been considered as a model for an intricate altitudinal zonation of climate, vegetation and agriculture. In addition to this, latitude, proximity to the Pacific Ocean or Amazon Basin, topography, hydrology and geomorphology enrich the variety of landscapes. In terms of agriculture and rural development, a corollary of other factors shapes the human landscape. Particularly significant among them are the ethnic affiliation of the population with their cultural heritage, the colonial and post-colonial imprint, land tenure, accessibility to roads and larger settlements, agricultural and non-agricultural opportunities, the access to and acceptance of innovations and modernization, and also the resilient capability of the rural population to adapt to climate change and to new cultural, social, economic, and political conditions. This chapter attempts to explore, in a summarizing fashion, the agricultural and rural archipelagoes of the tropical Andes in their horizontal and vertical dimensions. In a concluding part, the author critically examines some rural scenarios and postulates a "*campesino*-oriented development".

Keywords: tropical Andes, rural and agricultural archipelago, horizontal and vertical zonation of landscapes, *campesino*-oriented development

> *The most profound meaning of the Andes comes not from a physical description, but from the cultural outcome of 10 millenia of knowing, using, and transforming the varied environments of western South America ([1]: 34).*

1. Introductory remarks

Tropical mountain environments can be approached in a three-dimensional perspective taking into consideration the horizontal or lateral as well as the vertical dimensions of geographical space: Zimmerer [2] speaks of "vertical environments". In the case of the tropical Andes, the configuration of the natural environments and of the human landscape is further differentiated by the extent of the Cordilleras on both sides of the Equator from the Caribbean coast (about 11°N) to the tropic of Capricorn (about 23.4°S). Facing the Pacific Ocean with its different ocean currents on its western side and the vast interior, lowland areas of the Orinoco, Rio Negro and Amazon watersheds to the east furthermore result in a marked landscape contrast as one crosses the mountain ranges and highland basins from west to east.

As early as 1807, von Humboldt and Bonpland described the vertical arrangement of ecological zones in their famous illustration of climate and vegetation of the Chimborazo in Ecuador [3]. Troll [4, 5] and Lauer [6–8] described and compared the altitudinal zonation of climatic factors and vegetation in tropical mountains in general and also specifically in the Andes. They distinguished the principal zones of the *tierra caliente*, the *tierra templada*, the *tierra helada* and the *tierra nival* or *nevada* from the base to the top of high tropical Andean mountains. They further differentiated between the humid, semi-humid, semiarid and arid Andes and illustrated these zones by their famous three-dimensional altitudinal and latitudinal models. They also showed that the climatic characteristics of the tropical Andes have a major impact on land use, settlements and agricultural activities. Of great significance are in particular critical temperature thresholds, e.g. for the growth of specific tropical cultigens and of the occurrence of frost. In terms of humidity levels, the humid and semi-humid Andes are characterized by between 12 and 7 humid months (precipitation higher than potential evaporation), the arid and semiarid Andes by 6 to 12 arid months (evaporation higher than precipitation). In a generalized model, the author attempted to portray the altitudinal zonation of ecology, agricultural land use, settlements, and health risks for the humid and the semiarid and arid Andes (**Figure 1**).

A pioneering contribution to the concept of altitudinal ecological and human zonation was made by Murra [9, 10]. He states that life of the rural Andean world was shaped by the "verticality" of ecological conditions and that families, villages and ethnic communities have traditionally attempted to control as many micro-ecological zones as possible (*Control Vertical* or *Mitimagkuna*), the so-called *archipiélagos verticales*. Drawing on Murra's work and based on his own research, Brush [11–13] distinguished three major types of control and integration of Andean ecological zones and resource areas. The "compact type" is one in which different ecological zones occur in close proximity to each other and are easily accessible to the community. In the case of the "archipelago type", the ecological zones used by a group of peasants are more distant from each other and are often separated by unused areas, thus requiring more extended travel times. This may require the establishment of a series of permanent or semipermanent "colonies", away from the home community, in these different ecological zones, as well as a system of exchanges between the home community and the colonies based on reciprocity and redistribution. In the "extended type", each peasant group exploits a single or a few ecological zones, often specializing in certain products, and exchanges goods with other groups living and exploiting other ecozones ([11]: 292-295). In a summarizing overview, Forman [14] has discussed the "verticality concept" with its implications and applications for the Andes. She comes to the conclusion that the verticality models still provide useful guidelines for rural development in the Andes.

In a rather provocative paper, Allan [15] had rejected the "environmental determinism" of traditional altitudinal zonation models, arguing that they are "no longer suitable for characterizing mountain ecosystems now that human activity is directed to new motorized transportation networks linked to a wider political economy and no longer dependent on altitude" ([15]: Abstract, 185). Instead, he proposed an "accessibility model" of land use in a hypothetical mountain landscape. While mountain geographers would agree that a simplistic and unrestricted environmental determinism has to be rejected, many of them (among them [16]: 197-198), based on their empirical findings, have taken the position that mountain people for a long time have adapted to the geofactors of altitude, relief, distance, climate, vegetation, soil and hazard exposure, while recognizing that new developments, among

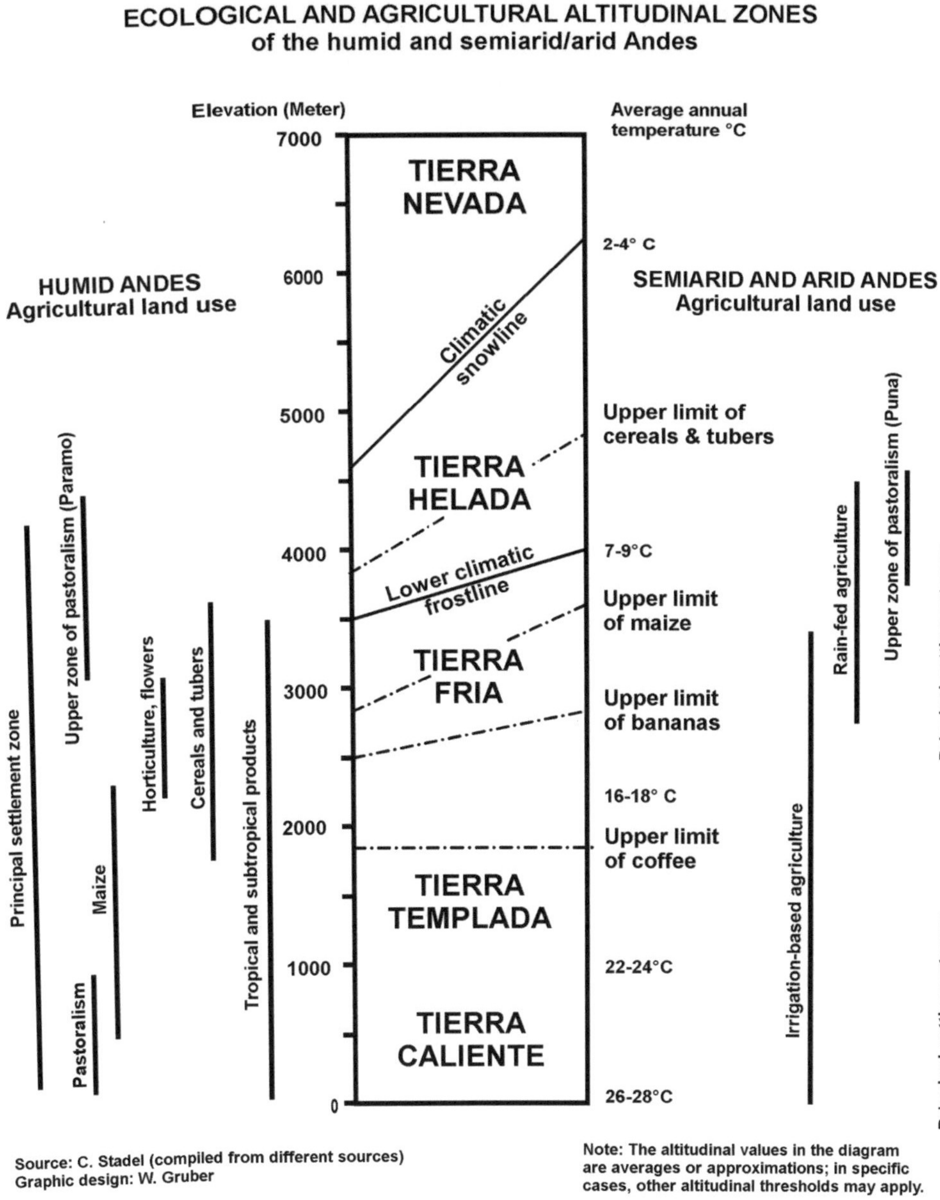

Figure 1.
Altitudinal zonation of ecology and agricultural land use in the tropical Andes (Stadel 1989).

them accessibility, transportation and intensified lowland-highland interactions, have influenced and modified human activities in mountains. In his rural research in Ecuador, Peru and Bolivia, the author [17–20] identified a vast array of factors influencing agricultural activities and rural land use:

- Altitude and relief configuration, erosion and sedimentation.
- Distance, proximity or remoteness to service centres and core areas.
- Climate, vegetation and soils.
- Natural hazards.

- Conservation measures.
- Access to and distribution of water: precipitation regimes, water rights and irrigation schemes.
- Cultural and spiritual traditions and local perceptions and practices.
- Age and nature of settlement process.
- Population parameters: age and gender structures and mobility and migration.
- Land tenure, land ownership, water rights and land reforms.
- Access to acceptance of innovation, modernization and new technologies.
- "Conscientization" levels, education and training.
- Local, regional, national and global market conditions.
- Alternative economic activities and employment opportunities.
- Access to capital and investment opportunities.
- Transportation and communication and social infrastructures.
- Local leadership and community initiatives.
- Exogenous impact of business ventures, governmental programmes, non-governmental intervention and influences of "expatriates" (e.g. remittances, investments).

2. Horizontal and vertical agricultural and rural spaces in the tropical Andes

For a long time, agriculture has been the backbone of the rural economy and employment and has been the basis for ancient civilizations in the tropical Andes. Andean agriculture is characterized by a great variety of production systems, land-use forms, types of cultivated plants and domestic animals and forms of pastoralism. Due to the constraints of altitude, slope, climate, soil, forest cover in humid parts and barriers of difficult accessibility, only a limited part of the Andean realm is suitable for agriculture. The agricultural core areas are situated in the larger longitudinal and transverse valleys (e.g. the valleys of the Magdalena and Cauca rivers in Colombia; the Patate-Pastaza rivers in Ecuador; the Marañon, Santa Marta and Mantaro rivers in Peru; the Rio Grande in Bolivia; the Central Valley in Chile; as well as the river oases of the semiarid and arid of the Pacific realm in Peru and northern Chile). Other favored agricultural regions are the highland basins (e.g. in the Sabana of Bogotá), the *cuencas* or *hoyas* in Ecuador and the wide *Altiplano* in southern Peru and Bolivia, especially the shores of Lake Titicaca with their favorable microclimate. In addition, the inner flanks of the *Cordilleras* in the climatic zones of the *tierra templada* and *tierra fría* are intensively used agrarian regions. In contrast to the old settled and agriculturally used Andean realm, newer agrarian colonization zones and rural pioneer spaces have emerged at the eastern Cordilleran flanks and valleys in Colombia, Ecuador, Peru and Bolivia. Nonagriculturally oriented core areas are the urban–rural continuum regions of the major cities and metropolises, as well as the larger mining zones and manufacturing districts. Furthermore, population concentrations have developed along major transportation corridors and around principal ports and airports (**Figure 2**).

Andean agriculture is characterized by a pronounced altitudinal zonation, a result of thermic, hygric and edaphic differentiation. Following the classical altitudinal ecological "belts" from the warm lowlands to the cold highest parts of the ecumene, the *tierra caliente*, *tierra templada*, *tierra fría* and *tierra helada*, Borsdorf and Stadel [21] distinguish the following major agrarian zones:

1. Tropical lowland rain-fed farming (*Campo de Lluvia*) in the *tierra caliente* (from sea level to about 1000 m in the humid Andes)

2. Tropical lowland irrigation farming (*Campo de Riego*) in the *tierra caliente* in the semiarid and arid realm

3. Extratropical agrarian foothill zones (foremost the Chilean longitudinal valley and the foothill regions in northwestern Argentina)

Figure 2.
Chimborazo region, Ecuador (Photo credit: Stadel).

4. Agrarian areas of the tropical Andes in the altitudinal zones of the *tierra templada* and *tierra fría* (about 1000 to 4000 m)

5. Upper zones of field cultivation and pastoralism in the *tierra helada* (approximately 4000 to nearly 5000 m)

It is evident that water supply, water rights, water use and the management of the water resources are crucial for agriculture and rural sustainable livelihoods. Permanent, periodic or seasonal water scarcity and the high demand and diverse use of Andean water resources by a variety of decision-makers and often conflicting interest groups make water a critical ecological, cultural, economic, social and political issue and challenge. For instance, the excessive water consumption of the irrigated plantations of export-oriented river oases of coastal Peru threatens the water supplies for small-scale farming and rural communities in the upper watersheds. A voracious consumer of water is the powerful mining sector with its dramatic impact on the natural environment, the ensuing critical shortage and the contamination of water in the surrounding rural areas and the landscape degradation. More recently, the water demands in major tourist destinations (e.g. the Cordillera Blanca region, Cuzco and the *Valle Sagrado* in Peru) may conflict with the interests of farmers and rural residents in these areas. Conflicts may also arise in the use of water between the upper and lower parts of watersheds, between *indígenas* and non-native regions, between *latifundistas* and *minifundistas* and between urban and rural areas.

In a detailed study of a landscape profile of the Ecuadorian Sierra, Stadel [17, 18] investigated the complex ecological, agricultural and rural mosaic from the upper limit of agricultural activities and settlement at the foot of Chimborazo (about 4200 m) through the high mountain basin (*Cuenca*) of the city of Ambato and the Patate and Pastaza valleys to the foothills of the Eastern Cordillera (about 900 m). Along this altitudinal profile, the following land-use zones can be identified (**Figure 3**):

1. The sparsely settled pasture regions of the cool humid *páramo* at the upper limit of sporadic settlement and patchy niche field cultivation 3200 to 4200 m). The mostly indigenous population suffers from climatic stress and poor access to the market centres of Ambato and Guaranda; however, the *indígenas* control a large part of the regional water resources.

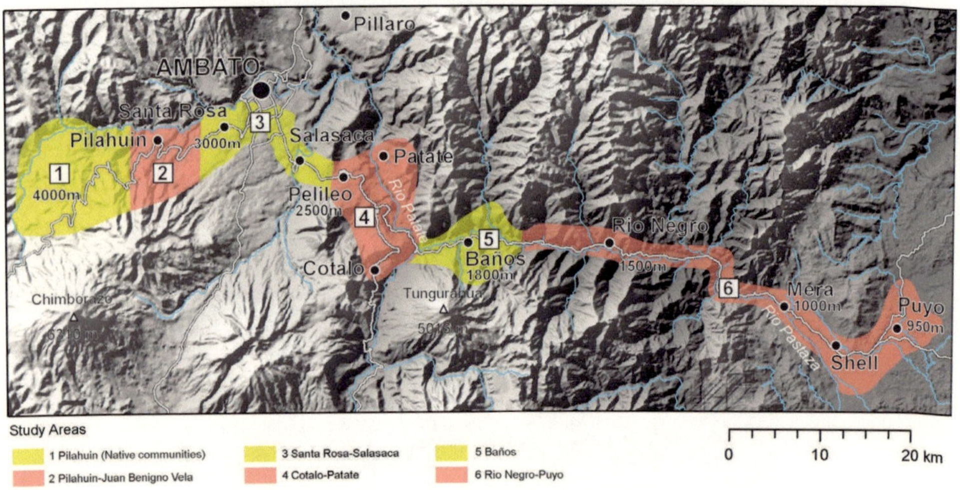

Figure 3.
Study region Chimborazo – Puyo, Ecuador (Stadel 1989).

DOI: http://dx.doi.org/10.5772/intechopen.86841

2. The upper zone of intensive arable farming (2800 to 3200 m). A vast array of crops is cultivated, mostly in seasonal or annual rotation. In the lower parts, precipitations tend to be insufficient and unreliable, and irrigation becomes necessary.

3. The high mountain basin (*Cuenca*) of Ambato, including the adjoining inner Cordilleran slopes (2500 to 2800 m), a mixed urban–rural space. Rural population clusters are located around the dynamic regional market centre of Ambato and are specializing in productive fruit and vegetable growing and also rely on job opportunities in the city. The climate is semiarid, and agriculture depends on irrigation. In the southeastern part, the small enclave of the Salasaca native community gives the cultural landscape a distinct identity.

4. The agricultural core region along the Patate valley and a major highway corridor to the *Oriente*, the gateway to the Amazon lowlands (2000 to 2800 m). In the deeply entrenched valley floor, a highly productive irrigation-based *hacienda*—and *minifundio*—agriculture contrasts with a mixture of irrigated and nonirrigated small fields. Here, a mixture of vegetables, cereals and fodder crops is grown in a variety of traditional rotation cycles on steep slopes. Above about 2600 m, the irrigation-based agriculture gives way to a mostly seasonal and rain-fed agriculture. The urban centre of Pelileo, located on the major highway to Ambato and the *Oriente*, is the principal market centre of the region and a new centre of textile manufacturing, especially a production of jeans for national and international markets.

5. The temperate humid part of the Pastaza valley (1200 to 2000 m). This section is located in the ecological zone of the *tierra templada* and benefits from the rains which reach this valley from the Amazon lowlands. In the narrow valley floor and lower slopes, a variety of subtropical and tropical fruits and vegetables are grown. In the higher reaches, a mixture of different cops of a temperature, cooler climate. As one proceeds further downstream, the steep slopes are increasingly covered with a dense humid montane forest. The centre of this section is Baños, a regional service centre, a popular site for Ecuadorians and also foreign visitors, as a pilgrimage site and a recreational destination because of its mild climate and thermal waters.

6. The lowest part of the landscape profile, located in the *tierra templada* and higher parts of the *tierra caliente* (900 to 1200 m). This is a permanently warm and very humid zone, characterized by recent colonization agriculture, and a dispersed linear pioneer settlement stretching along the highway. Here, a wide selection of tropical crops is grown in the valley and on patchy forest clearings on the mountain slopes. At the exit of the Pastaza from the Cordillera, the city of Puyo is the booming regional multifunctional centre.

Zimmerer [22] has pointed out that "overlapping patchworks of farm special units are characteristic of the mountain landscapes of Andean regions of Peru and Bolivia. Patchiness and overlap...are shaped by the broad tolerances of major crops, high variability/low predictability of habitat factors, multifaceted cropping rationales of cultivators including their linkages to extraregional influences, and, to varying extents, the sociospatial coordination of crop choice among farmers". Zimmerer arrived at this conclusion from detailed field studies within the two communities of Pampa Churigua (farmland range 2800 to 3450 m) in the Department of Cochabamba, Bolivia, and of Mollomarca (farmland range 3100 to 4100 m) in the Cuzco Department of Peru. Although a maize/cereal zone of the lower slopes can be distinguished from an upper potato/tuber zone), a considerable mixing of a variety of crops, a patchiness of land parcels and an elevation-related overlap of crop types can be observed. In another contribution, Zimmerer [23]) states that "integrating the conservation of biodiversity by smallholder farmers with agricultural intensification is increasingly recognized as a leading priority of sustainability and food security amid global environmental and socioeconomic change". This will contribute to an in situ conservation of agrobiodiversity and enhance the smallholders' resilience.

The traditional pattern of agricultural land use has been profoundly altered in some areas by the locational influences of accessibility to highway arteries and regional market centres (**Figure 4**). Where topography, soil quality and irrigation potential exist, a specialized cultivation of vegetables, fruit and flowers serves the urban market, in some cases even international markets (e.g. the plantation of cut flowers for global markets in the *Sabana de Bogotá*) (**Figure 5**). Other agricultural cores of a specialized, export-oriented agriculture have developed because of an early valorization of favorable ecological conditions (e.g. the coffee-growing zones of the *tierra templada* in Colombia), or they have been the result of modernization, new technologies and entrepreneurial initiatives (e.g. the cultivation of special vegetables such as asparagus for world markets in the river oases of coastal Peru). Other important specialized agricultural zones are the wine-growing areas of the Central Valley of Chile and of the Cuyo region of Argentina or the legal or illegal plantation of coca bushes on the humid eastern side of the Andes in Bolivia, Peru and Colombia. New consumer demands may also entail a specialization of agricultural strategies. Examples for this are the new

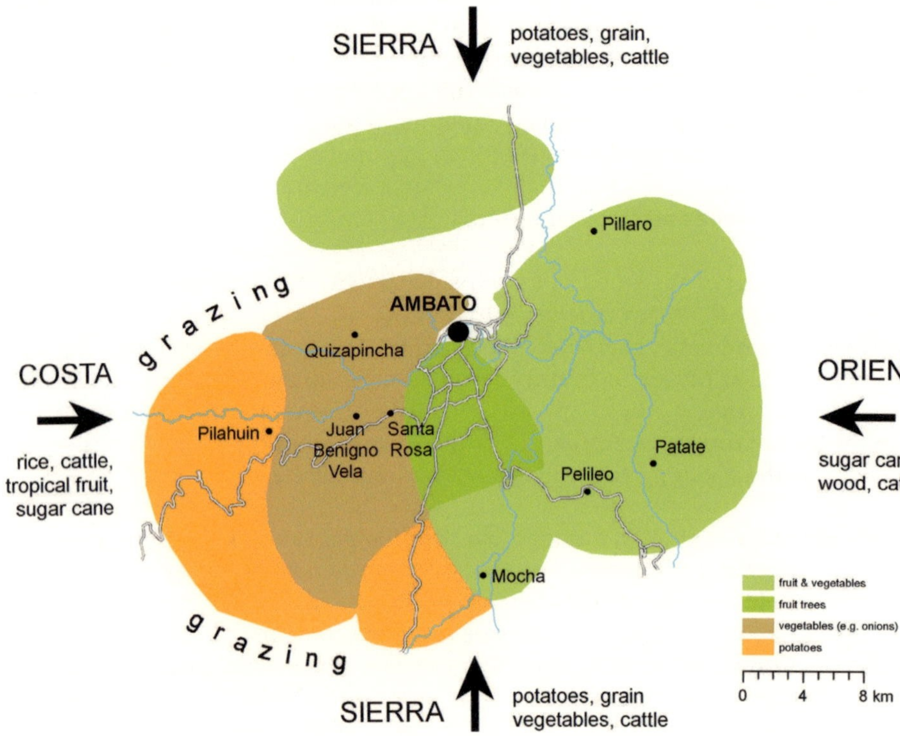

Figure 4.
Ambato market centre and agricultural hinterland, Ecuador (Borsdorf and Stadel 2015).

DOI: http://dx.doi.org/10.5772/intechopen.86841

Figure 5.
Greenhouses of commercial flower cultivation, Sabana of Bogotá, Colombia (Photo credit: Stadel).

quinoa monocultures in the Lake Titicaca region or expanding alpaca breeding on the Bolivian *Altiplano*. While this specialization may bring enhanced economic benefits to the region, the potentially negative impact on the ecology, regional water resources, land tenure, traditional land use practices and potentially higher farming risks cannot be ignored [19].

3. Peri-urban clusters

Metropolitan centres and other important regional capital centres and economic centres have experienced major population growth rates and areal expansions. This has resulted in a massive planned or uncontrolled urban–rural interface of a wider surrounding region and to the emergence of major peri-urban clusters ([21]: 184-188 and 191-192). While this urbanization may bring to the region new housing, attractive landscape amenity sites for affluent urbanites (so-called *parcelas de agrado* and *ciudades valladas*, **Figure 6**), new employment opportunities or enhanced infrastructures, the negative impacts of this "urban invasion" often prevail ([24]: 239). Land speculation and soaring land prices are threatening the survival of small-scale agriculture and the traditional rural livelihoods by a consumption of often fertile irrigated agricultural land and by diverting the water resources from irrigating the fields to a use for urban households and commercial needs. Driven away by this urbanization process, agricultural smallholders are faced with the options of incorporating themselves into the urban agglomeration, to intensifying land use on their remaining plots or to seeking alternative new agricultural areas. Haller [24] has found that farmers in the Huancayo basin have expanded or intensified field cultivation in the higher *suni* [25] altitudinal belt (3500 to 4000 m), a marginal and poorly accessible agricultural zone with steep and nonirrigated slopes not suitable for year-round cultivation. Using the example of the regional city of Huancayo and the lower Shullcas Valley, Haller and Córdova-Aguilar [26] have demonstrated that urbanization puts pressure on agrarian land use, endangers the environmental integrity of the region and impacts the Huaytapallana Regional Conservation Area.

In the Andes, these agglomerations of a dynamic and multifunctional urban–rural continuum represent the most important areas of population growth, land use

Figure 6.
Ciudad Vallada Piedra Roja, Chile (Borsdorf and Stadel 2015).

conversion and excessive densities of buildings and infrastructural developments. These newly emerging or rapidly expanding clusters are facing the challenge of integrated and effective regional planning and policy actions that attempt to regulate the nature of the growth processes, to recognize the interests of urban and rural stakeholders and to harmonize economic goals with ecosystem services.

4. Mining clusters

Since early times, mining has played a major role in the economic development of the Andes. With the discovery of rich deposits of the precious ores of gold and silver, mining has resulted in the establishment of working camps and subsequently in the foundation of smaller and larger settlements. The most famous of them were the silver-mining city of Potosí (**Figure 7**) in current Bolivia and the mercury-producing city of Huancavelica in the Peruvian Sierra. Both of these booming centres of the early colonial mining industry are located at high altitudes, Potosí at close to 4100 m and Huancavelica at 3600 m. After the initial and generally short-lived

Figure 7.
Potosí, Bolivia (Photo credit: Stadel).

gold- and silver-mining boom, other mineral deposits became important: copper, tin, zinc, lead, iron ore, salpetre and most recently lithium.

Unlike farming and agricultural settlements, the development of mining settlements was not related to favorable environmental factors; many mining sites emerged in locations normally considered unfit for settlements: copper mining in the arid Atacama desert and the mining of gold and a range of non-precious ores at high elevations, some of them above the limit of the ecumene of farming and pastoralism. The most striking example for this is La Rinconada in the southern Peruvian Andes, a gold-mining boom town at 5100 m with an estimated population of some 40,000 people. In addition, many mining clusters developed in areas of poor accessibility and the building of adequate transportation lines represented a major challenge. While the development of these mining areas largely superseded environmental constraints, mining and the associated smelting activities had entailed a corollary of environmental impacts, not only for the mining settlements proper but also for a larger surrounding region, e.g. the excessive consumption of regional water resources, deforestation, severe erosion, mass wasting processes and water and air pollution.

While mining may offer to the regional population often a much needed alternative employment, encourage the development of infrastructures and services and have stimulated regional economies, the mining sector for a long time has been controlled and dominated by outside national and foreign stakeholders who had little interest in a sustainable regional development. Bury [27] portrayed the negative repercussions of mining on traditional land tenure, water rights, agricultural land use and community institutions. Furthermore, the fate of mining tends to be fluid and uncertain, with many mining areas affected by the typical "boom and bust cycles" resulting from an exhaustion of ores or sharply declining global market prices.

5. Rural tourism nodes

"The exceptional diversity of landscapes and cultures in the Andes holds rich opportunities for tourism" ([21]: 249). The ecological variety in the tropical and extratropical realm of the Andes ranges from the humid rainforests (*selva*) and cloud forests (*Ceja de la Montaña*) to various types of highland grasslands, to thorn steppes, saltpans (*salares*) and deserts. On the highest summits in the tropics and on lower elevations in the extratropical realm, snow- and icefields cover the mountains. In addition to this extraordinary ecological diversity, the impressive mountain scenery of rugged peaks (most famous of them are the Torres del Paine in Chilean Patagonia or the spectacular, glacier-covered Cordillera Blanca in Peru) of the numerous active and dormant volcanoes; the deeply entrenched valleys (e.g. the Colca Canyon in Peru); the impressive fjord coast of southern Peru, the impressive rivers in the eastern Cordilleras; the mountain lakes, foremost Lake Titicaca; or the vastness of the Peruvian and Bolivian *Altiplano*, the landscape appeal of the Andes is further complemented by the rich cultural heritage of the region. Among the famous visitor attractions are the pre-Inca sites (e.g. Chan Chan in the coastal desert of Peru; Chavín de Huantar in the eastern Cordilleras of Peru; Tiahuanaco on the Bolivian Altiplano); the impressive monuments and other relics of the material culture of the Incas centred in Cuzco, Machu Picchu (**Figure 8**), and other sites of the *Valle Sagrado*. With the Spanish conquest, the Inca culture was superseded and replaced by the Spanish cultural heritage. Visitors are attracted to the colonial towns with their churches, monasteries, museums, *plazas* and typical colonial houses, to attractive *hacienda* buildings (many of them converted to elegant hotels) and to

Figure 8.
Machu Picchu, Peru (Photo credit: Borsdorf and Stadel).

pilgrimage sites. Many of the pre-Spanish and colonial cities have been included in the UNESCO World Cultural Heritage list, a fact which further enhances their appeal for tourists. In a generalized model, Borsdorf and Stadel [21] have portrayed the altitudinal zonation in major types of Andean tourism in the tropical and extratropical realm (**Figure 9**).

In addition to the most common type of a sightseeing tourism attracted to the most famous sites, other forms of visiting and tourism can be observed. Ecotourism has been promoted at all altitudinal levels in many ecologically attractive niches. Of particular interest to visitors are the National Parks, the Biosphere Reserves and other types of protected areas. Still rather spotty are various forms of rural or agro-tourism, but this type of tourism may still face the barriers of difficult accessibility, substandard accommodation and other facilities, insufficient investment funds and promotion and sometimes also hesitant rural host families and communities. Successful examples are the *comunidades* of Vicos and Humacchuco in the Cordillera Blanca region of Peru, to the north of the major mountain tourist centre of Huaraz [28]. With the support of the *Instituto de Montañas* in Huaraz, the local population was involved in various ways in a gentle, ecologically and culturally compatible and sustainable rural tourism.

Under the motto "cuidar la vida en las montañas" (protecting life in the mountains), some communities around the Huascaran National Park (founded in 1975) benefit from this initiative and are participating in all stages of the planning and management of rural tourism. Ecotourism and "soft" agrotourism are contrasting with newer forms of sports, adventure or event tourism (e.g. mountain biking, paragliding, white-water rafting, modern festivals). Mountaineering, here called *andinismo*, has a long tradition and appeals to a national and international clientele. Preferred destinations are the high Cordilleras, notably the Cordillera Blanca and Cordillera de Huaylas of Peru, the Cordillera Real in Bolivia and the Patagonian Cordillera of Argentina and Chile.

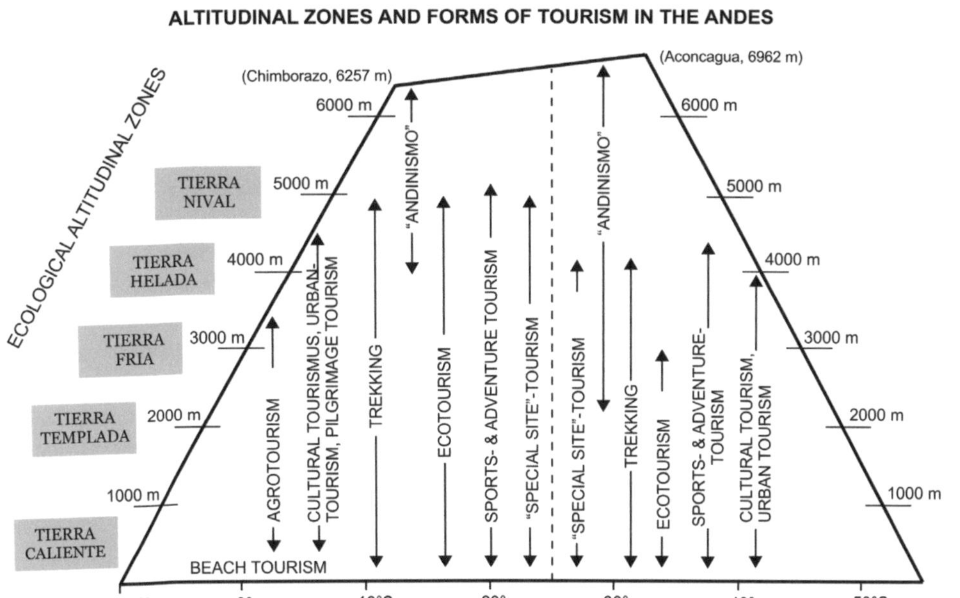

Figure 9.
Altitudinal zones and forms of Andean tourism (Borsdorf and Stadel 2015).

6. Rural spaces: development scenarios and options

With progressive urbanization, rural spaces have lost some of their former demographic weight and economic importance. Nevertheless, rural populations continue to represent a large share of the tropical Andean states, and the rural realm forms an important part of national identities and cultures. Economically, many areas can still be rated as marginal spaces, but many regions are important as diversified agrarian areas, as water reservoirs, as mining sites, as destinations for urban amenity migrants and tourists and most important as livelihoods for people. Some rural core areas have become new growth poles and arenas for development and modernization; other regions, in particular the poorly accessible and resource-deficient areas, are threatened by natural hazards, by poverty, stagnation and marginalization, aggravated by political, economic and social neglect and discrimination. External influences and impulses pervade the entire rural realm, even the remote areas. Today, electronic information and communication media bring rural people in touch with national and global developments. In addition, temporary or permanent out-migrants furnish their home community external information, in many cases also remittance cash flows or investments. This has a significant economic, social and cultural impact on their former home communities. Further external actors are government agencies, an array of non-governmental organizations, international institutions and powerful corporations and companies. The consequence of these impacts are significant "livelihood transitions" and "place transformations" [29] which may even transform some Andean core regions into globalized spaces [30].

The result of these multiple endogenous and exogenous influences may have positive or negative impacts on rural communities and livelihoods:

In some of the more accessible areas, technological innovations and market developments have stimulated agricultural developments and changes in crop patterns, leading to serious consequences for exchange relationships and trade between zones. In other zones, people have diversified their livelihood through non-agrarian activities (crafts, wage labour, etc.) or have migrated. ([31]:3).

Yarnall and Price [32] have examined the impacts of migration and remittance flows on communities in the Valle Alto of the Department of Cochabamba in Bolivia. They observed a "new rurality" transforming the traditional rural environment and society. The communities have benefited by being linked to new "diaspora knowledge networks", from increased material resources and new stimuli of development. Some formerly poor peasant communities have even become materially better off than nearby colonial towns. But at the same time, the remittance dependence has made these communities vulnerable; as for various reasons, these cash and investment flows may not be reliable and sustainable. Furthermore, traditional forms of agricultural activities and employment may be eroded, and emerging rather sharp economic and social disparities result in a fragmentation of the rural realm.

It follows that rural development is complex, highly differentiated and at times also controversial. A generally accepted approach is to harmonize environmental, sociocultural and economic goals. Bebbington [33] views rural development neither solely rooted in conventional cultural values, economic pursuits and social structures and to the persistence of a subsistence-based economy nor in an uncritical opening to external influences, modernization, new technologies and an unrestricted adherence to national and global market processes. An array of development interventions are directed toward an attenuation of natural risks and their impacts, a protection of natural resources and a preservation of the genetic pool of biodiversity. But these efforts can only have a long-term success if the livelihoods of rural communities and the basic needs of the local population are secured or enhanced in a sustainable fashion [34]. Furthermore, it is today generally recognized that local and regional cultural aspects should be the fundamental basis in the development discourse ([21]: 311). In the past, rural economic development was often guided by external views and strategies without considering the "meaning that campesinos impart to the economy as actors in a social context" ([35], Abstract, 310). Rist therefore pleads for an "actor-oriented approach that is not based on preconceived, nonlocal concepts" (ibid.). This has been referred to as "ethno-development" [36], "development with identity" [37], "participatory cultural development" and other terms. Andolina et al. [38] call this approach "alternative modernities" enabling and mobilizing local human resources and strengthening local ownership and responsibility [39]. Local cultures and the traditional heritage are no longer seen as obstacles and barriers to development, but as enriching, locally accepted and sustainable factors. Local knowledge and practices should not be seen as static and paralyzing, but as dynamic and evolving: "transformed by autochthonous innovations, by an adaptation to changing circumstances, and by an adoption of knowledge, capabilities and technologies" ([40]: 14, translated).

Based on his empirical research in the tropical Andes, Stadel [20] derived the following postulates for a "*campesino*-oriented development":

- Appreciation of the knowledge and experience of *campesinos* (*saber campesino*) and strengthening of their cultural pride.
- Esteem for the traditions, cultural values, customs and rituals of local communities (*lo andino*, [41]; *sagesse des Andes,* [42]).
- Strengthening of communal solidarity and cooperation.
- Respect for nature (*cosmovisión andina*) and an aspiration to harmonize environment and society.
- Exploration of the potentials and limitations of the natural and human environments.
- Strengthening of the resilience and adaptive capacities of the local population, facing environmental risks, economic and social vulnerabilities and potential disaster.
- Improvement of the living conditions of the population, with a special focus on poor people and enhancement of the infrastructures and services in water supply, sanitation, health, nutrition and housing.
- Promotion of environmentally compatible and sustainable forms of agriculture (*agroecología*) and silviculture and of agricultural niche products.
- Enhancement and diversification of alternative income and employment opportunities (e.g. in eco- or agrotourism).
- Mobilization of local human resources and creation of attractive local perspectives for young people to stem their migration to cities.
- Improved access to microloans and other forms of financial and technical support.

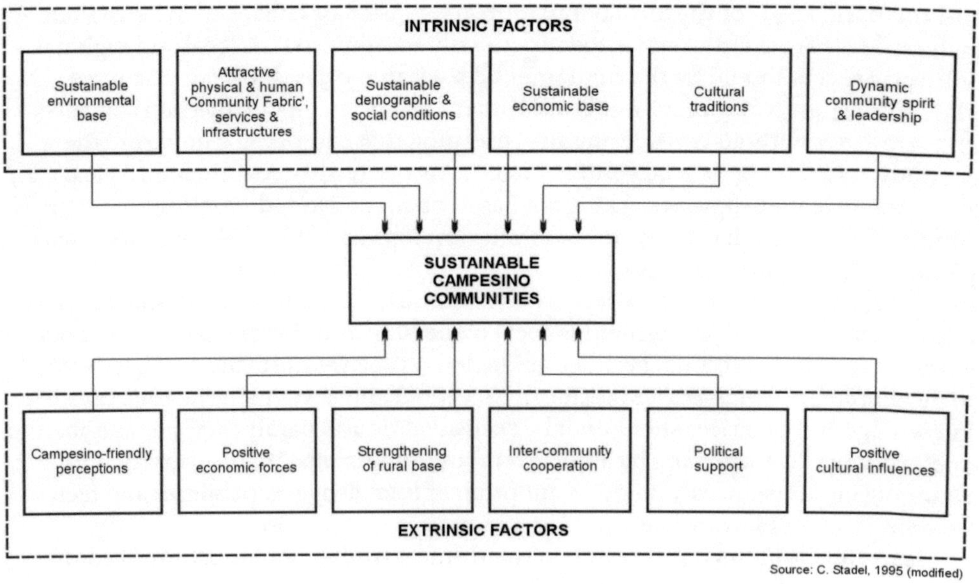

Figure 10.
Sustainable campesino communities – a conceptual model (Stadel 2008).

- Sensible use of external funds, especially of the remittances, to meaningful types of investment.
- Safeguards against economic, social and political discrimination and exclusion and struggles against external exploitation.
- Development emphasis on locally perceived and formulated needs, priorities and implementation methods.
- Participation, enablement and empowerment in rural development and ownership of projects by local communities.
- Enhanced communication channels, accessibility and transport facilities.
- Improvement of the quantity and quality of formal and informal education and training.

In a simplified summarizing table (**Figure 10**), Stadel [20] has proposed a conceptual model for "sustainable *campesino* communities". It is argued that *campesino* communities can benefit by various positive intrinsic factors, as well as by favorable extrinsic factors.

7. Conclusion

In spite of rapidly expanding metropolitan centres and a progressing urbanization, the identity of the Andean realm is still rooted in agricultural traditions and in rural societies. Based on the mountainous character but also because of the opportunities for rural living, the Andes can be portrayed as a rich and varied mosaic of agricultural fields, pastures, farms, villages and towns, forming archipelagos of favorable environmental conditions, of human activities, and of cultural heritages.

The diversity of rural spaces is the result of the extraordinary variety of natural and cultural traits, both in the horizontal and vertical dimensions of the Andes. In the horizontal perspective, agricultural land use in the tropical regions is distinguished from that of the extratropical one and is also differentiated by climatic influences from the Pacific Ocean or from the continental basins of the Amazon and Orinoco watersheds. Distinct agricultural patterns and rural landscapes are further resulting from the human factors of accessibility to roads and markets, cultural traditions, as well as external impacts.

While the core and most widespread functional identity of the Andes lies in farming, pastoralism and agricultural settlements, the rural space is also shaped by other activities, foremost mining, industry and commercial activities. More recently, urban real-estate interests have "invaded" selected regions outside larger cities, especially in areas with a specific landscape or climatic appeal. Urban "amenity migrants" have moved into secluded peri-urban clusters, often into "gated communities" (*ciudades valladas*). Another newer form of rural functional orientation is the recreational appeal and the national and international tourism potential in attractive landscapes and cultural sites. Therefore, the extraordinary complexity of micro-spatial rural clusters has generated an intricate pattern of diverse "archipelagos" in the Andes.

The rural Andes are a dynamic realm undergoing many changes and deep transformations. This applies to agriculture with its adaptation to changing environmental conditions, to new market orientations and in some cases to altered perceptions and strategies of farmers. Rural regions, even in formerly remote locations, are no longer isolated areas; in some cases, they may also no longer be regarded as peripheral spaces. New transportation arteries and communication channels connect rural residents to national core areas, even to global regions and actors. But the changes in the rural realm have not eliminated its disparities, and the "new rurality" has old and new winners and losers. Some regions are stagnating, and some rural people remain poor or are becoming marginalized, while others are dynamic, with its stakeholders progressing and seizing new opportunities.

The viability of the rural Andes is endangered by a number of internal and external threats. The vagaries of the climate and environmental deterioration processes are threatening agriculturally based livelihoods, especially those of small farmers. The persistent imbalance in the land tenure system, rural unemployment and underemployment, poverty and deficient infrastructures and services, combined with the lure of cities and other countries, have depleted many rural regions of the human capital of young and enterprising people. Furthermore, the growing external control of the land and its natural resources by external interests and stakeholders threaten the livelihoods of the rural population.

What are the options for a sustainable future of the rural Andes? Generally speaking, the rural realm must be effectively assisted to overcome inequality, discrimination, poverty and marginality and thus become an attractive living space and an alternative to the life in large cities or overseas. Rural population should be empowered to control and mobilize their resources and to develop mechanisms for enhanced local autonomy and self-determination. The author has proposed a generalized conceptual model for "sustainable *campesino* communities". But every region and community has its own identity, needs and priorities and will undoubtedly find their ways to enable them to seek appropriate development paths, likely in a careful balance between proven environmentally and culturally adapted strategies and new ones, innovative but also sensitive to the environment, societies and cultures of the region: "The pursuit of sustainability is a local undertaking not only because each community is ecologically and culturally unique but also its citizens have specific place-based needs and requirements" ([37]: 1).

Author details

Christoph Stadel
University of Salzburg, Austria

*Address all correspondence to: christoph.stadel@sbg.ac.at

References

[1] Gade D. Nature and Culture in the Andes. Madison: The University of Wisconsin Press; 1999

[2] Zimmerer KS. Vertical environments: Science, nature, and cultural-environment diversity in modern development. In: Offen K, Dym L, editors. Mapping Latin America. Chicago: The University of Chicago Press; 2011. pp. 263-268

[3] Humboldt von A, Bonpland A. 1807, reprinted in 1959: Essai sur la Gégraphie des Planètes. In: Accompagné d'un Tableau Physique des Régions Equinoxiales. Paris: Cotta (1807). London: Society for the Bibliography of Natural History; 1959

[4] Troll C. Die Tropischen Gebirge. Ihre Dreidimensionale Klimatische und Pflanzengeographische Zonierung. Bonner Geographische Abhandlungen 25. 1959

[5] Troll C. Geo-Ecology of the Mountainous Regions of the Tropical Americas. Bonn: Verlag Ferdinand Dümmler; 1968

[6] Lauer W. Vom Wesen der Tropen. Klimaökologische Studien zum Inhalt und zur Abgrenzung eines Irdischen Landschaftsgürtels. Wiesbaden: Akademie der Wissenschaften Mainz; 1975

[7] Lauer W. Natural potential and the land-use system of the Kallawayain the upper Charanzani valley (Bolivia). Erdwissenschaftliche Forschung. 1984;**18**:173-196

[8] Lauer W. Geoökologische grundlagen andiner agrarsysteme. Tübinger Geographische Studien. 1987;**96**:51-71

[9] Murra JV. El control vertical de un máximo de pisos ecológicos en la economía de las sociedades andinas. In: Formaciones Económicas y Políticas del Mundo Andino. Lima: Instituto de Estudios Peruanos; 1975. pp. 59-115

[10] Murra JV. The limits and limitations of the 'vertical archipelago' in the Andes. In: Masuda S et al., editors. Andean Ecology and Civilization. Tokyo: University of Tokyo Press; 1985. pp. 15-20

[11] Brush SB. El lugar del hombre en el ecosistema Andino. Revista del Museo Nacional (Lima). 1974;**40**:277-299

[12] Brush SB. Mountain, Field, and Family: The Economy and Human Ecology of an Andean Valley. Philadelphia: University of Pennsylvania Press; 1977

[13] Brush SB. The natural and human environment of the Central Andes. Mountain Research and Development. 1982;**2**(1):19-38

[14] Forman SH. The future value of the 'verticality' concept: Implications and possible application in the Andes. In: Allan NJR, Knapp G, Stadel C, editors. Human Impact on Mountains. Totowa: Rowman and Littlefield; 1988. pp. 133-153

[15] Allan NJR. Accessibility and altitudinal zonation models of mountains. Mountain Research and Development. 1986;**6**(3):185-194

[16] Uhlig H. Do accessibility models make altitudinal zonation models obsolete? Mountain Research and Development. 1986;**6**(3):197-198

[17] Stadel C. Del valle al monte. Landnutzung und höhengliederung im raum patate-pelileio. Die Erde. 1975;**116**(1):7-25

[18] Stadel C. The perception of stress by Campesinos—A profile from the ecuadorian sierra. Mountain Research and Development. 1989;**9**(1):35-49

[19] Stadel C. L'agriculture andine: Traditions et mutations. In: Crises et Mutations des Agricultures de Montagnes. Clermont-Ferrand: CERAMAC; 2003. pp. 193-207

[20] Stadel C. Agrarian diversity, resilience and adaptation of Andean agriculture and rural communities. Colloquium Geographicum. 2008;**31**:73-88

[21] Borsdorf A, Stadel C. The Andes. A Geographical Portrait. Cham et al.: Springer; 2015

[22] Zimmerer KS. Overlapping patchworks of mountain agriculture in Peru and Bolivia: Toward a regional-global landscape model. Human Ecology. 1999;**27**(1):135-165

[23] Zimmerer KS. The compatibility of agricultural intensification in a global hotspot of smallholder agrobiodiversity (Bolivia). Proceedings of the National Academy of Sciences. 2013;**110**(8):2769-2774

[24] Haller A. The "sowing of concrete": Peri-urban smallholder perceptions of rural-urban land change in the central Peruvian Andes. Land Use Policy. 2014;**38**:239-247

[25] Pulgar Vidal J. Geografía del Perú: Las Ocho Regions Naturales, la Regionalización Transversal, la Sabiduría Ecológica Tradicional. 3rd ed. Lima: PEISA; 1996

[26] Haller A, Córdova-Aguilar H. Urbanization and the advent of regional conservation : Huancayo and the cordillera Huaytapallana, Peru. Economics Monthly. 2018;**10**(2):59-63

[27] Bury J. Mining mountains: Neoliberalism, land tenure, livelihoods in the new Peruvian mining industry in Cajamarca. Environment and Planning A. 2005;**37**(2):221-239

[28] Torres AM. Participatory planning for ecotourism development in the Peruvian Andes. Yale F & S Bulletin. 1996;**99**:284-294

[29] Bebbington A. Reencountering development: Livelihood transitions and place transformations in the Andes. Annals of the Association of American Geographers. 2000;**90**(3):495-520

[30] Bebbington A. Globalized Andes? Livelihoods, landscapes, and development. Ecumene. 2001;**8**(4):414-436

[31] Salman T, Zoomers A, editors. Imaging the Andes: Shifting Margins of a Marginal World. Amsterdam: Aksant; 2003

[32] Yarnall L, Price M. Migration. Development and a new rurality in the Valle Alto, Bolivia. Journal of Latin American Geography. 2010;**9**(1):107-124

[33] Bebbington A. Social capital and rural intensification: Local organizations and islands of sustainability in the rural Andes. The Geographical Journal. 1997;**163**(2):189-197

[34] Zoomers A. Linking Livelihood Strategies to Development. Amsterdam: KIT publishers; 1999

[35] Rist S. Linking ethics and the market: Campesino economic strategies in the Bolivian Andes. Mountain Research and Development. 2000;**20**(4):310-315

[36] Laurie N, Andolina R, Radcliffe SA. Ethnodevelopment: Social movements, creating experts, and professionalizing indigenous knowledge in Ecuador. Antipode. 2005;**37**:470-495

[37] Rhoades RE, editor. Development with Identity. Community, Culture, and Sustainability in the Andes. Oxfordshire: CABI Publ; 2006

[38] Andolina R et al. Indigenous Development in the Andes: Culture, Power, and Transnationalism. Durham: Duke University Press; 2009

[39] Gibbs D. Exploring local capacities for sustainable development. Geoforum. 2005;**36**:407-409

[40] Neubert D, Macamo E. Entwicklungsstrategien zwischen lokalem Wissen und globaler Wissenschaft. Geographische Rundschau. 2002;**54**(10):12-17

[41] Stadel C. Lo Andino': Andine umwelt, philosophie und weisheit. Innsbrucker Geographische Studien. 2001;**32**:143-154

[42] Regalski PA. La Sagesse des Andes. Une Expérience Originale dans les Communautés andines de Bolivie. Geneva: Fondations Simón L. Patiño & Pro Bolivia; 1994